Title.

The Atmospheric Sciences and Man's Needs

PRIORITIES FOR THE FUTURE

Committee on Atmospheric Sciences
NATIONAL RESEARCH COUNCIL

NATIONAL ACADEMY OF SCIENCES
WASHINGTON, D.C. 1971

ISBN 0-309-01912-5

Library of Congress Catalog Card Number 70-611003

First printing, May 1971
Second printing, December 1971

Available from

Printing and Publishing Office
National Academy of Sciences
2101 Constitution Avenue
Washington, D.C. 20418

PREFACE

This report identifies ways in which the atmospheric sciences can contribute to important human needs, and it proposes an ordering of research priorities necessary to meet these needs. Clearly, in preparing a report of this scope, competence covering science, engineering, public policy, and education is essential. The study plan leading to this report was designed to tap the broad range of expertise required and to encourage expression of all points of view, while ensuring that the report could move forward to completion and review on a reasonable schedule.

Responsibility for organizing the study and for determining conclusions and recommendations was vested in a Steering Committee. Eleven subject-matter panels covering scientific and technical disciplines were charged with (a) reviewing scientific and technical problems, constraints, recent achievements, and existing research plans and (b) discussing the contribution of each subject area to the fulfillment of specific human needs. Individuals were assigned responsibility to review national investments in scientific and technical resources in the categories of facilities, manpower, government research, and university and institutional research. These panels and individuals provided the major input to the study. For each of the areas of major application—prediction, atmospheric quality, modification, and information services—an individual was assigned the responsibility to assemble, summarize, and interpret the relevant material from the panels and resource persons. Participants met for two weeks at the Friday Harbor Laboratories of the University of Washington to discuss panel reports and to help to formulate recommendations. A Study Director was appointed to coordinate the various activities

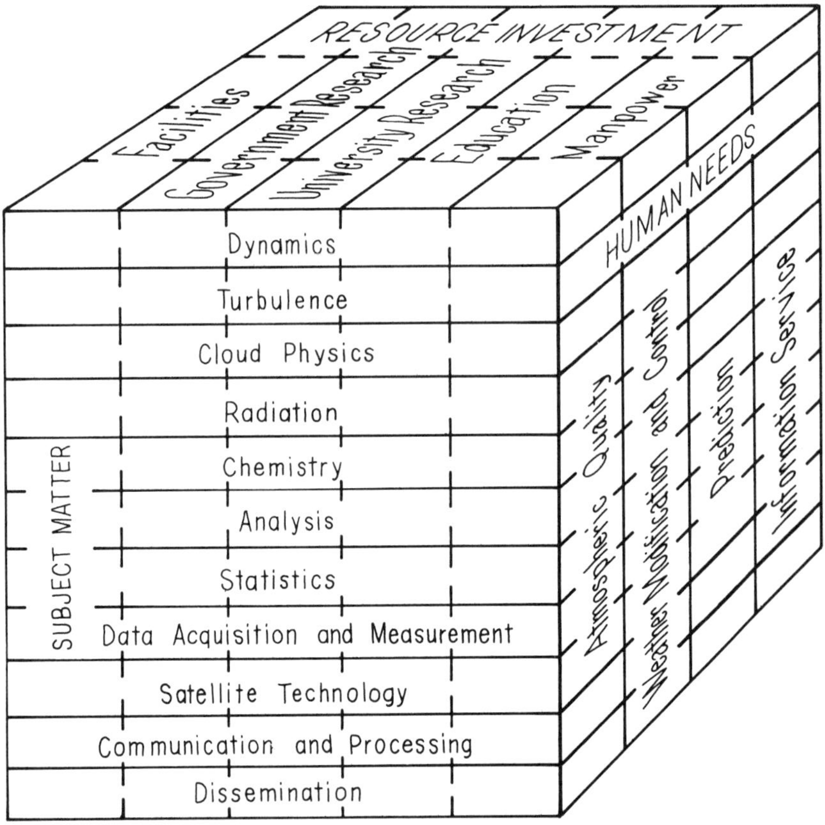

and to edit the report. More than 55 individuals have participated actively in the study. University department chairmen, government officials, and committees of the American Meteorological Society were invited to comment on particular aspects of the study, and many of these have provided useful perspectives.

The figure above shows the relationships of various inputs to the study. The matrix may be viewed as a data-processing and communications device, providing both access to a large store of information and judgment and internal channels for feedback, for computing, and for selective retrieval.

The major conclusion and recommendations of the report represent a broad consensus. With few exceptions, the difference of views expressed by the participants has been limited to matters of detail or emphasis. This is of special significance because the recommended actions imply substantial shifts in some instances from past patterns of research effort.

The study was supported by the following government agencies: the Department of Commerce; the Department of Health, Education, and Welfare; the Department of the Interior; the Department of Transportation; the Atomic Energy Commission; the National Aeronautics and Space Administration; and the National Science Foundation.

ROBERT G. FLEAGLE, *Chairman*
Committee on Atmospheric Sciences

CONTENTS

COMMITTEE ON ATMOSPHERIC SCIENCES

Chairman	ROBERT G. FLEAGLE University of Washington
Vice-Chairman	C. GORDON LITTLE National Oceanic and Atmospheric Administration
	LOUIS J. BATTAN University of Arizona
	HENRY G. BOOKER University of California at San Diego
	GEORGE F. CARRIER Harvard University
	JULE G. CHARNEY Massachusetts Institute of Technology
	WILLIAM W. KELLOGG National Center for Atmospheric Research
	PAUL E. KLOPSTEG Glenview, Illinois
	THOMAS F. MALONE University of Connecticut
	WILLIAM A. NIERENBERG Scripps Institution of Oceanography
	VERNER E. SUOMI University of Wisconsin
	EDWARD TELLER University of California at Livermore
	PHILIP D. THOMPSON National Center for Atmospheric Research
Executive Secretary	JOHN R. SIEVERS

1
INTRODUCTION

For scientists, as for us all, the proper study of mankind is man.

SIR ROBERT WATSON WATT

PURPOSE

The decade of the 1960's has seen a strengthening of the research and operational capabilities of the atmospheric sciences, with a concomitant increase in capabilities for useful application on a scale heretofore impossible. Simultaneously, environmental problems have come forcefully to public attention. We begin the decade of the 1970's, therefore, with a unique combination of opportunities and challenges—a new capacity for useful applications to the needs of society at the same time that the power of public opinion provides a driving force toward solution of pressing problems. These circumstances are likely to have far-reaching influences on the atmospheric sciences in coming years. For these reasons, it seemed essential in this report to move beyond the limits of the scientist's traditional interest and to consider science in its relation to human needs.

Discussions with government science administrators have emphasized to the Committee that determination of priorities for investment in atmospheric research and its application has become extremely difficult. Resources are not adequate to support all scientifically valid and useful programs. There are growing gaps between scientific proposals and resources, between the public's rising expectations and the limited capabilities that inevitably have resulted from five years of essentially level funding. We believe that in the next decade the atmospheric sciences must continue to attack fundamental scientific problems, and that at the same time more attention should be directed toward human needs. Priorities will have to be determined. As scientists we cannot expect to have the sole voice in setting priorities—nor should we. We do have the responsibility for scientific assessment of possible applications and for proposing a basis for rational ordering of priorities in the atmospheric sciences.

In light of these considerations, the Committee has undertaken in this report to identify from the many possible applications of the atmospheric sciences those that can contribute most to human needs within the next five to ten years, to suggest the broad outlines of realistic programs to meet these needs, and to arrange the recommended actions according to priority.

In identifying the human needs discussed in this report, the Committee recognizes that thorough assessments of the benefits of weather forecasting or of other weather services do not exist; that dissemination of information to the public is based largely on traditional procedures using outmoded technology; and that there is often insufficient interaction between the user and the information system. For these reasons, we have been somewhat handicapped in carrying out our designated task. Nevertheless, we consider as obvious the great value to society of intelligent control of air pollution, timely warnings or modification of dangerous weather events, and the growing economic and social importance of accurate forecasts.

An alternate objective might have been to determine priorities on the basis of intrinsic scientific interest. This would have been less responsive to current needs and also would have been much more difficult because the purpose of basic science is to discover the unknown, to open new possibilities for further research and for applications. The past record of science amply demonstrates the wisdom of supporting basic research, but it provides little guidance for the establishment of priorities for the future. Therefore, the priorities discussed in this report are for the most part concerned with near-term (five to ten years) applications of the atmospheric sciences to human needs. Longer-term applications can be expected to emerge from the recommended research programs and also from research motivated more simply on the basis of intrinsic scientific interest.

The choice of objectives has the effect that several fields

within the atmospheric sciences are given nonuniform attention. The field of *meteorology* is represented much more strongly than is *aeronomy* or *planetary atmospheres*, to use the three categories followed in the budgetary breakdown by the Interdepartmental Committee on Atmospheric Sciences (ICAS). Whereas meteorology has more important near-term applications at this time, the applications of aeronomy may be of longer term. In discussing this matter, the Committee also noted that broad and comprehensive reviews of aeronomy have been carried out in the past three years by the National Academy of Sciences–National Research Council[1,2] and that a review of planetary atmospheres is included in the current study by the Space Science Board.[3] There should be no question of the intrinsic scientific importance of aeronomy and planetary atmospheres, because of their role in our efforts to understand the evolution of the solar system and the interaction of the atmosphere and the sun.

The choice of objective also has the effect that certain important areas within meteorology are discussed explicitly, while others—for example, applications of meteorology to agriculture, aviation, and marine operations—are not. This report is not intended as a summary review of the state of the science, as was carried out by the Committee ten years ago in the Petterssen-Little report,[4] nor is it a comprehensive analysis of scientific or technical subject matter. Such comprehensive technical reviews have been carried out in recent years for the fields of air–sea interaction, global prediction, weather modification, and remote sensing.[5–8]

THE NATURE OF ATMOSPHERIC RESEARCH

Atmospheric scientists are concerned with understanding how the complex fluid-dynamical system that is the atmosphere works; for example, with understanding how storms are produced, how pollutants are removed from the atmosphere, and how the ocean and atmosphere interact. These are a few of the basic scientific problems that occupy many of our best theoreticians. At the same time, the solution of each of these problems can be applied directly to immediate practical problems. Therefore, research in the atmospheric sciences often is at the same time basic and applied, and the distinction between basic and applied is more a reflection of individual interest and motivation than of the scientific nature of the problem or the mode of research. The importance of basic understanding is clearly evident in virtually all applications of the atmospheric sciences. Consequently, major applications should be based on strong basic-research programs.

A more important distinction exists between individual research and group research. The dominant atmospheric problems, whether they be basic science or applications, require group efforts using substantial resources. This is a consequence of the large extent of the atmosphere and the fact that many phenomena are strongly influenced by a large number of interacting processes, e.g., surface properties, aerosol characteristics, insolation, evaporation, and the presence or absence of clouds. To understand atmospheric processes in anything approaching a general sense and to apply understanding in useful ways require group research and large programs. For this reason, this report is particularly concerned with problems requiring group efforts for their solution.

New ideas, however, are essentially individual. In order to provide the milieu that is needed for new ideas, discriminating support must be provided for undirected individual research by the brightest and most creative scientists. There is no other way to keep the total research enterprise productive and healthy. One of the most important challenges facing research management is to devise effective means for including these individual scientists and small groups in large group-research programs directed toward major goals. The leadership of creative scientists of broad understanding is indispensable.

THE 1960's: A DECADE OF PROGRESS

Introduction

The atmospheric sciences today are hardly recognizable as descendants of their ancestors of 40 years ago, the descriptive meteorology and climatology of the pre-World War II era. However, in that now distant period, concepts and theories were developed which, together with the great expansion in upper-air observations and in professional manpower that occurred during the war, laid the basis for a postwar period of active research. Major attention from the end of the war to the late 1950's was directed toward understanding the mechanism of the general circulation and toward development of numerical models that simulated large-scale atmospheric features. In this period, the

Air Force was the leading federal agency in supporting research. Following the launch of Sputnik in 1957, the atmospheric sciences expanded on a broad front. The fields of aeronomy and planetary atmospheres developed rapidly, and new emphases were placed on meteorological research in cloud physics and chemistry, turbulence and air–sea interaction, geophysical fluid dynamics, and the technology of global observations. Thirteen federal agencies have invested in atmospheric research and development over the past decade. The National Aeronautics and Space Administration (NASA) and the Air Force provided the major support for aeronomy and planetary sciences, while NASA, the National Science Foundation, and the Department of Commerce have become leading supporters for meteorological research and development. This period of broad scientific effort has continued throughout the 1960's, and significant advances in the science and technology of the atmosphere have occurred in a variety of fields. Understanding of the atmosphere has increased rapidly, and the gap between science and useful application has been substantially reduced in many areas. Among the recent advances the following have been especially significant.

Atmospheric Dynamics

Basic understanding of the global circulation was achieved some 20 years ago, and this achievement has led to development of a series of increasingly detailed and accurate numerical prediction models. The following significant intellectual steps have marked the course of this development: the linear theory of baroclinic instability, mathematical filtering of unwanted acoustic and gravity waves, development of essentially realistic numerical models of the general circulation, inclusion of inertia–gravity effects in the models, discovery of finite difference algorithms capable of eliminating computational instability, development of models capable of simulating present climate, and development of the models used today in operational weather forecasting. This sequence provides a fascinating chronicle of science deserving of a fuller recount than is appropriate here.

More recent theoretical developments have provided important new insights into atmospheric circulations; these may in the future contribute to further applications of dynamics to weather prediction and modification. The problem of atmospheric tides has at last been solved,[9] and the role of gravity waves in vertical propagation of momentum has been elucidated.[10]

Important advances have been made in the understanding of large-scale tropical phenomena. Numerical simulation of the development and life cycle of hurricanes has progressed markedly, although the models remain circularly symmetrical.[11,12] Linear models of the tropical atmosphere have been shown to admit combined gravity–Rossby waves which resemble observed waveform disturbances.[13,14]

The fundamental predictability of the atmosphere has been examined, and provocative questions have been raised.[15–17] Inherent instabilities, both physical and computational, limit the range of deterministic prediction, probably to a few weeks or less.[18] However, there may be orderly aspects of atmospheric processes that permit prediction of certain statistics for greater periods.[19]

Global Observation and Prediction

In the past decade, development of the first generation of satellites and sensor systems has made possible observation of the global atmosphere. Numerical simulation of the atmosphere has kept pace with theory and with observing capability.[18,20] At the same time, development of computer technology has enabled us to process the enormous amount of data that describe the atmosphere. The next generation of computers should at last make feasible the regular prediction of the global circulation using mathematical models that include the major physical processes occurring in the atmosphere. The atmospheric sciences have until now been data-limited, but this restriction can be overcome. These technological developments make a great difference in the class of problems that are amenable to study.

The Dryden-Blagonravov agreements of 1962 and 1963 marked the recognition that satellite technology provided the capability for a great advance in atmospheric observation and in prediction. During the succeeding years a great deal of effort in this country has been directed toward global observations and prediction. International plans have been instituted under the Global Atmospheric Research Program (GARP) and the World Weather Watch (WWW) to extend the range of useful weather prediction to probable limits of 1 to 2 weeks.* The National Academy of Sciences has issued two reports setting forth the scientific justification and the need for specific re-

*GARP Publication Series, 6 vols. (1969–1970), and GARP Special Reports, 2 vols. (1970), International Council of Scientific Unions–World Meteorological Organization, Geneva, Switzerland. World Weather Watch, Planning Reports, World Meteorological Organization, Geneva, Switzerland.

search and development in support of the GARP objectives.[21] The U.S. Committee for the GARP was organized in 1968 within the National Academy of Sciences to provide scientific guidance to our national efforts. The GARP and WWW, which together constitute the World Weather Program (WWP), have been endorsed by the President of the United States, and the 90th Congress has approved Senate Concurrent Resolution 67 which gives "full support to the world weather program." The National Oceanic and Atmospheric Administration (NOAA) is the designated lead agency for the program, and interagency coordination and guidance is provided by the Interdepartmental Committee for the World Weather Program (ICWWP). The United States is now well embarked on a variety of scientific and technological subprograms that will contribute essential parts to the global program. Global weather prediction is an idea whose time has come.

The World Weather Program represents a fortunate confluence of scientific understanding and technical capability; it is directed to man's need to minimize the dangers of severe storms and to improve man's capability to plan the many activities that are strongly influenced by atmospheric changes. The global program also will provide the essential foundation on which other applications to be discussed in this report are based.

Remote Sensing

The atmospheric sciences are, of course, based on accurate observations. Traditionally, these observations have been made using sensors placed in the region to be observed. To extend such direct measurements from the relatively limited areas where they are now employed to the entire atmosphere, as required for global prediction, would be prohibitively expensive. Observations made with direct *in situ* sensors also may not be representative of the large areas or time periods desired. Furthermore, they offer insufficient resolution for observing small-scale phenomena. For these reasons, techniques of remote sensing are of great importance to the atmospheric sciences. These techniques provide high resolution, they integrate over time or space directly, and they permit direct determination of certain statistical properties of the atmosphere.

The past decade has been a period of rapid and significant advances in the theory and technology of remote sensing. The most dramatic examples have been the development and successful operation of satellite sensors. Optical and infrared cameras carried by polar-orbiting TIROS and geostationary applications technology satellites (ATS) have provided global cloud images which have been directly responsible for new insights into the scales of motion occurring in the atmosphere. Tropical cloud systems have been recognized as having unexpected coherence and organization,[22,23] and observations of cloud motions promise to provide the first nearly global measurements of wind velocities.[24] Most important, however, has been the success of the Satellite Infrared Spectrometer (SIRS) in providing accurate temperature soundings in regions of clear or partly cloudy sky.[25] Radiometric sounding from satellites may provide the necessary global data more economically and much sooner than had been anticipated a few years ago. These technical developments constitute critical contributions to the World Weather Program. They have resulted from the financial support and the combined planning and management skill of the National Aeronautics and Space Administration and the National Environmental Satellite Service.

Other remote-sensing instruments have been developed for ground-based and aircraft use. Doppler radar has been used to examine the turbulent motions occurring in precipitation clouds.[26] Small-scale structure has been detected in clear air in exquisite detail using highly sensitive microwave radar.[27,28]

Optical radar (lidar) equipment is being developed for the measurement of the distribution of water vapor and aerosol.[29] Acoustic echo-sounding can be used for remote measurement of wind and turbulence by utilizing the naturally occurring small-scale temperature fluctuations as atmospheric tracers.[30] This technique is especially suited to the study of the structure and dynamics of the boundary layer. A recent study emphasizes the great potential of remote-sensing techniques for making measurements on a wide variety of phenomena.[8] These techniques are destined to play a central role in each of the applications that are discussed later in this report.

Boundary-Layer Transfer

Most of the energy that drives the global circulation is provided to the atmosphere by evaporation from the ocean surface. Water vapor is transferred upward through the lower part of the atmosphere (the planetary boundary layer) and condenses in the cloud layer with consequent release of latent heat. These processes occur at a rate such that the residence time for water vapor in the atmosphere is about ten days. Therefore, to extend prediction to periods of ten days requires that the processes of evaporation at the surface, vertical transfer through the planetary boundary layer, condensation, and precipitation be incorporated into the global prediction model.

The mechanisms responsible for vertical transfer of water vapor, heat, and momentum through the planetary boundary layer have been the focus of considerable research effort during the decade. A 1962 study[5] emphasized the importance of research on interaction of the sea and the atmosphere in order to gain a better understanding of the dominant processes responsible for putting energy into the atmosphere. A variety of different research programs have been directed toward this objective. Substantial progress has been made in relating surface fluxes to properties of the free atmosphere,[31] and a fairly general empirical scheme for incorporating boundary-layer fluxes in general circulation models has been proposed.[32] Methods of measurement of boundary-layer properties have been developed for use on aircraft and on floating platforms. Several field experiments have been carried out, culminating in 1969 in the Barbados Oceanographic and Meteorological Experiment (BOMEX), operationally the largest and most complex field research program yet undertaken by atmospheric scientists and oceanographers. Analysis of the BOMEX observations is not complete, but early results in which aircraft and buoy measurements were compared indicate greater complexity of the structure and processes within the planetary boundary layer than has been understood previously.[33] We have tools far better than those available 10 years ago, and we have gained a wealth of valuable empirical information about the mechanisms of energy transfer. We have found, however, no easy route to fundamental understanding of the structure of the planetary boundary layer.

Significant advances have been made recently in understanding the linkage of the planetary boundary layer to the lower troposphere. The interaction of Ekman "pumping" produced by surface friction and condensation above the boundary layer has been proposed as an important mechanism in the development of hurricanes from small tropical disturbances.[34] The theory of Ekman instability has been shown to account for some of the secondary circulations occurring in the boundary layer,[35-37] but satellite photographs of cloud systems have revealed a variety of orderly circulation systems, which have not yet been explained. The Global Atlantic Tropical Experiment (GATE), planned for 1974, should substantially extend understanding of these systems.

Observations of coherent large-scale cloud systems in tropical regions have stimulated theoretical studies that have successfully related the boundary-layer convergence of the intertropical convergence zone to equatorial waves of a particular band of frequencies.[38-40] A theory of large-scale tropical circulation strongly involving the planetary boundary layer appears to be in the making.

Weather Modification

Laboratory and field experimentation have helped to clarify some of the problems associated with efforts to modify weather phenomena beneficially. Research programs, supported by various government agencies, have been devoted to the study of precipitation modification, the dissipation of fogs over airports, the reduction of the frequency of cloud-to-ground lightning and hail, and the mitigation of the effects of severe storms.

A review of the field of weather modification in 1966[6] was cautiously optimistic about various aspects of this field, particularly the dissipation of fogs and precipitation augmentation. Subsequent research in the United States and in other countries, reviewed by the NRC Panel on Weather and Climate Modification[41] indicates that significant progress is being made. The evidence is strong that in some situations precipitation can be increased, while in others it may be decreased. It now is necessary to learn how to distinguish the meteorological conditions under which increases or decreases occur.

Advances over the last few years offer encouragement that it should be possible to develop schemes for the clearing of warm fogs and the reduction of the frequency of cloud-to-ground lightning and of damaging hail and violent storms. The experimental and the theoretical evidence that it may be possible to reduce the strength of hurricane winds by means of cloud seeding is especially noteworthy.

Progress in the development of techniques for purposefully changing the weather has been accompanied by a growing realization that human activities have been inadvertently influencing the climate. The extent of the influence still needs to be determined.

Organizational Developments

During the past decade the following organizational developments have notably strengthened the capability of the atmospheric sciences to contribute to human needs.

University Research and Education The number of universities with major academic commitments to doctoral programs in the atmospheric sciences has doubled in the past decade, and the number of doctorates granted per year has more than doubled. Approximately half of the research publications of the decade were published by university faculty members and students. Nearly half of the regular academic faculties at the ten leading universities in the atmospheric sciences received their doctoral degrees within the past decade. These younger faculty members already are leading the way in many of the more sophisticated research fields.

These facts are evidence that university education and research in the atmospheric sciences has been substantially invigorated and strengthened in the past decade, and that the universities provide a large and essential part of the nation's scientific capability in the atmospheric sciences. The quality and vitality of university faculties and students provide assurance that the research programs proposed in the report can be competently carried out.

The growing strength of university programs has occurred with support from a considerable number of federal agencies. Especially effective has been the support provided by the National Science Foundation for research programs, facilities development, and fellowships and traineeships. The National Science Foundation's role, in the future as in the past, will be crucial to the continuing health and strength of the science and essential for the provision of well-trained manpower.

The National Center for Atmospheric Research The National Center for Atmospheric Research (NCAR) was created in 1960 with the primary role of developing research programs and coordinating university efforts concerned with crucial atmospheric problems too large to be handled by individual universities. This role has been demonstrated in several programs, for example, the organization and management of the 1967 Line Islands Experiment, the National Hail Research Experiment to be carried out during the next 6 years, and the Operational Systems Simulation Experiment which has been initiated in support of the GARP.

In addition, NCAR has developed strong research groups in atmospheric chemistry, turbulence, numerical modeling, cloud physics, and in other fields essential to the objectives of this report. NCAR by recruiting physicists, mathematicians, chemists, and engineers has materially enriched the research capability of the atmospheric-sciences community. It has served as a conference center and as a laboratory for graduate-student and faculty research. It has provided postdoctoral research experience and a wide variety of contacts with scientists from all over the world. In these ways, NCAR is a valuable national resource.

Environmental Science Services Administration and National Oceanic and Atmospheric Administration Several administrative steps have been taken by the government in the past 6 years to strengthen the atmospheric sciences and to coordinate more effectively the programs of the separate agencies. In 1965, the Weather Bureau, Coast and Geodetic Survey, and Central Radio Propagation Laboratory of the National Bureau of Standards were combined within the Department of Commerce to form the Environmental Science Services Administration (ESSA). In 1970, consolidation of the environmental sciences moved another step forward with the establishment of the National Oceanic and Atmospheric Administration (NOAA) within the Department of Commerce. The former scientific laboratories of ESSA now form the nucleus of atmospheric research strength for NOAA and have become leading centers in a wide range of scientific and technical areas: satellite instrumentation and interpretation of data, geophysical fluid dynamics, electromagnetic and acoustic-wave propagation, severe storms, and others. In 1969, ESSA provided the management and scientific coordination for BOMEX.

Creation of NOAA provides broad scientific and operational capability for marine and atmospheric investigations. It brings under central direction and funding the atmospheric and marine activities of ESSA; the buoy development program from the Department of Transportation; the Sea Grant Program from the National Science Foundation; elements of the Bureau of Commercial Fisheries, of the Bureau of Sport Fisheries and Wildlife, and of the Bureau of Mines from the Department of the Interior; the Great Lakes Survey from the Corps of Engineers; and the National Oceanographic Data Center and National Oceanographic Instrumentation Center from the Navy. Among its responsibilities is the monitoring of global air quality. NOAA employs about 13,000 persons, and the level of expenditures for fiscal year 1971 is about $310 million.

Environmental Protection Agency The independent Environmental Protection Agency (EPA) was created in 1970 to assume responsibility for control of air and water pollution and disposal of solid wastes. It combines the former National Air Pollution Control Administration, the Bureau of Solid Wastes Management, the Bureau of Water Hygiene, and portions of the Bureau of Radiological Health and pesticide aspects of the Food and Drug Administration, all from the Department of Health, Education, and Welfare; pesticide regulations from the Department of Agriculture; the Federal Water Quality Administration, the Gulf Breeze Biological Laboratory, and the Water Resources Division of the Department of the Interior; and certain responsibilities of the Atomic Energy Commission and the Federal Radiation Council. The EPA employs about 6000 persons. Most of EPA's initial budget (fiscal year 1971) of about $1.4 billion is allocated for grants for sewage treatment. With policy guidance from the Council on Environmental Quality of the Executive Office of the President, EPA should provide for the first time unified management and control of environmental quality.

PUBLIC CONCERN FOR THE ENVIRONMENT

The atmosphere, as the sustainer of life, is man's most vital resource. Fortunately, it is a self-renewing resource. The oxygen we use is ample in supply and is continuously replaced through photosynthesis, while natural processes normally act to cleanse the atmosphere of undesirable gases and particles. However, in the past decade we have had to recognize that the environment can be seriously and dangerously impaired by steadily increasing industrial production and individual consumption. Man-made pollution has become of increasing concern to large numbers of people, and possible dangers are discussed vigorously by scientists and the general public. Young people, especially, are scornful of a society that seems to place affluence above environmental quality.

Robert M. White expressed these concerns at the dedication of the ESSA Geophysical Fluid Dynamics Laboratory at Princeton on January 25, 1969, in the words: "Since time immemorial, we have been trying to protect man against his environment. Now we must try equally hard to protect the environment against man."[42]

Atmospheric scientists, along with other professional groups concerned with the environment, will be called on in the next decade for answers to difficult and far-ranging questions. Many of the questions already enuciated go well beyond our present capacity to answer accurately or thoroughly.

Such technical questions as the following must be answered:

1. What is the characteristic history (chemical reactions and movement) of various constituents of the atmospheric aerosol?
2. How will solar radiation, cloudiness, rainfall, and temperature be affected by a doubling of the rate of burning of fossil fuels in the next 20 years? Are there critical longer-term effects?
3. How can hurricanes or thunderstorms be safely reduced in severity?
4. How can weather prediction be utilized to maximize productivity for various industries (e.g., transportation, logging, agriculture, smelting) within constraints imposed by considerations of public welfare?
5. How can weather observations and predictions contribute to intelligent management of water resources and to the minimization of damage and loss due to floods and water pollution?

Questions such as these are difficult to answer and will demand considerable scientific effort; consequently, it will be essential to continue to emphasize quality in research and in training of specialists. But the resolution of environmental problems will not coincide exactly or even approximately with solutions to specific scientific problems. Practical decisions concerning environmental problems will be strongly influenced by social and political factors. Demand for more effective environmental management is likely to grow, perhaps enormously, over the next two decades. Are the atmospheric sciences, along with other environmental sciences, prepared to contribute responsibly to the creation of an environmental management "industry?" The dimensions of this question go well beyond the professional. Mechanisms of effective technological assessment will be needed which are responsive to the broad needs of society. There will be need to create a citizenry broadly informed and able to think effectively about environmental problems. The current growing interest in these problems should be a great asset to their solution. However, environmental problems are complex, and one should not expect them to yield easily. Society will have to learn that everyone cannot have both high-performance automobiles and clean air. And serious problems of conflict of interest must be resolved if weather modification is to be applied widely.

These considerations have implications for education at all levels from grade school to graduate school. College students already have taken a significant role in re-examining the traditional curriculum and in urging that new areas of study be created to handle problems of environmental quality. In the effort to solve problems whose dimensions involve the humanities as well as the natural and social sciences, these students propose to ignore departmental and discipline boundaries. Faculties tend to emphasize the necessity for training in a discipline if any headway is to be made on technically difficult problems. Resolution of this issue is by no means clear at this time. It is clear, however, that (a) solutions of environmental problems will require coordinated efforts on a broad intellectual base, (b) universities have the broadest range of capabilities available in any of our institutions, (c) many university students are highly motivated to attack complete environmental problems. Partially in response to this situation, profound changes are likely to occur in university structure and activities. We anticipate a considerably broader context for study of the atmosphere and at the same time increased interest in specialized study and research on atmospheric problems.

CONCLUSION

The growth of scientific capability has made possible increased useful application of the atmospheric sciences at the same time that certain environmental problems are approaching critical levels. Enlarged opportunities exist for extending our understanding of the atmosphere and for applying this understanding more effectively than ever before to human needs. We look forward to a decade of scientific challenge in which atmospheric scientists will assume increasing responsibilities in the developing arena of environmental management.

2 WEATHER PREDICTION

Perhaps someday in the dim future it will be possible to advance the computations faster than the weather advances and at a cost less than the saving to mankind due to the information gained.

L. F. RICHARDSON

The central function of the atmospheric sciences is to achieve the ability to predict changes in the atmosphere within useful limits of accuracy. The fundamental equations that describe the behavior of the atmosphere are time-dependent or predictive, and therefore, we can claim to understand an atmospheric phenomenon only when, in some sense, we are able to predict it. Alternately, virtually all important applications of the atmospheric sciences are based on prediction. Thus, it is appropriate first to point out some general considerations that influence our ability to predict atmospheric behavior before discussing applications.

Immediately, we are confronted with time and space scales ranging over eight or more orders of magnitude, from high-frequency turbulent fluctuations to changes in climate occurring over millenia. Phenomena of nearly every scale are important: for example, turbulent fluctuations are responsible for transfer and for dispersion of pollutants, storms provide cities with water, while climate changes may create a dust bowl or raise lake levels.

In Figure 1 are indicated the range of scales of atmospheric variability that are important to the topics discussed in this report. Some of the familiar weather phenomena with which man must contend

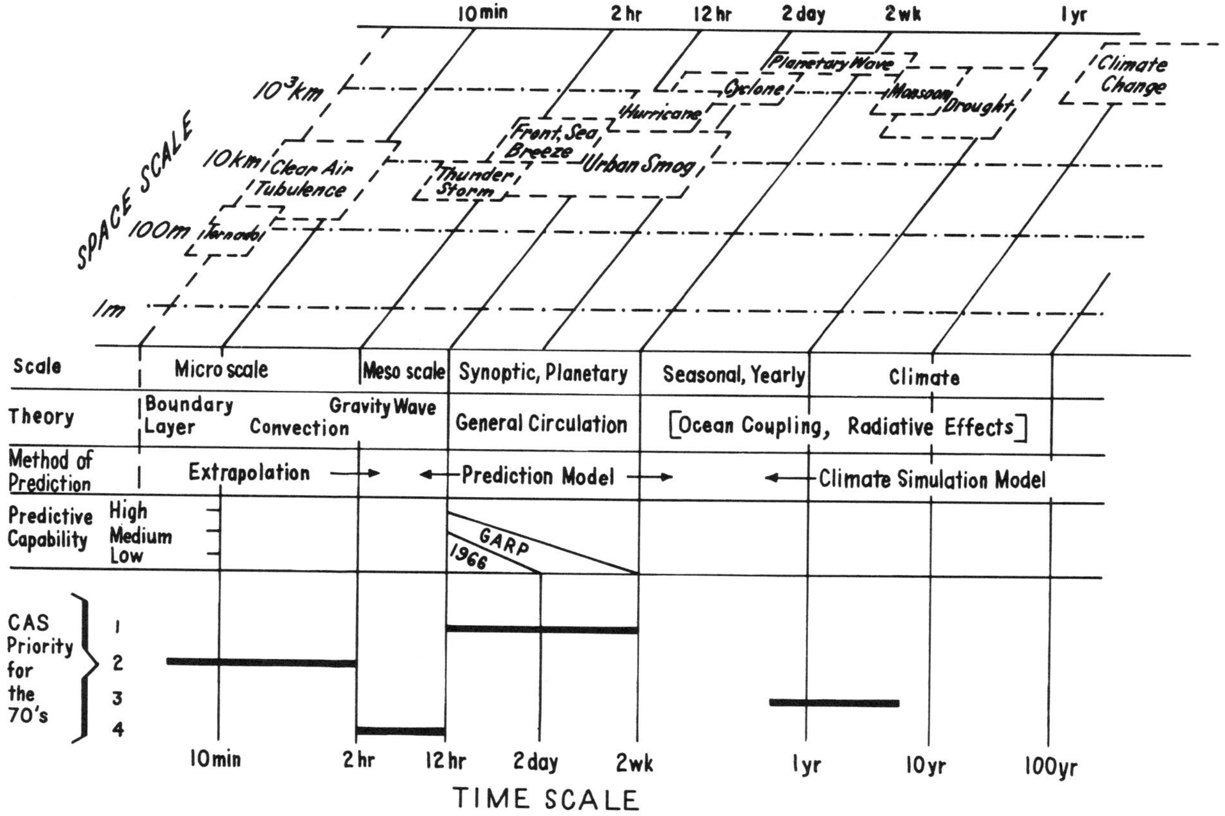

FIGURE 1 Time and space scales of atmospheric phenomena and their prediction.

are represented at their characteristic time and space scales. *Time scale* is used in this context to denote the dominant period of variations observed at a point, and *space scale* denotes the horizontal dimension of the various phenomena. Other phenomena could be added to Figure 1, but these examples demonstrate that important atmospheric phenomena occur over a wide range of time and space scales.

The lower part of Figure 1 identifies five time-scale intervals: the microscale, mesoscale, synoptic and planetary, seasonal and yearly, and climate. For each scale range the theoretical basis for our present understanding is indicated. Speculative theories are indicated in brackets. The line labeled "Method" indicates the method of prediction appropriate to each scale. The triangles opposite "Predictive Capability" indicate that routine operational predictive capability is claimed for the synoptic and planetary scales and that predictive capability for the other scales is lacking. Finally, the priorities recommended in this chapter are indicated schematically at the bottom of the diagram.

SYNOPTIC AND PLANETARY SCALES

Operational weather-prediction capabilities just a few years ago were limited for the most part to the time interval from 12 hours to 2 days, the synoptic and planetary scales as indicated by the triangle labeled "1966" in Figure 1. These capabilities were based largely on barotropic numerical model or, alternatively and more or less equivalently, on evolution of empirically determined forecasting rules. The numerical models in use prior to 1966 took no account of energy or mass input through the boundary layer, dissipation within the atmosphere, condensation and precipitation, radiation, or the influence of small-scale motions. In the past 5 years, operational forecasting centers as well as research institutes have developed more sophisticated models that account for certain of these effects, and significant improvements in forecast accuracy have been achieved.[43,44] The National Meteorological Center now issues regular 72-hour prognostic charts and is often successful in predicting new storm development. However, the lack of global data holds the period of useful forecasts to a limit of a few days. The World Weather Program is intended to extend this limit to something approaching the "theoretical limit" for deterministic synoptic scale models of about 2 weeks.*[,18,45] The area labeled "GARP" in Figure 1 indicates that as a result of these efforts predictive capability should be increased in the period between 12 hours and 2 weeks. Better forecasts of cyclones, planetary waves, hurricanes, and urban smog should result. Already progress has been made in acquiring global data and in developing numerical nodels, so that predictive capability now has been extended somewhat beyond 2 days.

On the short period or mesoscale side of the synoptic and planetary region of Figure 1, numerical prediction models using small grid scales have been applied to a minor extent to smaller-scale phenomena such as fronts. Progress has been limited due to the computational requirements and to the fact that vertical convection and condensation play such important parts in mesoscale phenomena. Although numerical models have successfully simulated convection cells, the dynamics and statistics of convection have not been incorporated in the models of the large-scale circulation; this constitutes one of the most serious limitations on progress in atmospheric prediction.

CHANGES OF MICROSCALE AND MESOSCALE

Despite the fact that little predictive capability can be claimed for time scales less than 12 h (Figure 1 indicates no capability), a significant fraction of atmospheric variability occurs on these smaller time and space scales. Tornadoes, thunderstorms, and other destructive storms are included here, but predictions are rarely sufficiently specific in terms of intensity, location, and time of occurrence. Terminal forecasts are made for aircraft, and probability forecasts are issued for severe storms expected to occur in large regions. Short-period severe storm warnings are issued in certain regions. Figure 1 indicates that there are theoretical bases for understanding at least some of the important phenomena. Why then has short-range prediction not advanced as rapidly as has prediction on the synoptic and planetary scales? Two factors explain the situation.

*This limit refers to the results of numerical experiments that show that in about 2 weeks of forecast time small errors in the initial data or subgrid scale features may grow until their effects are as large as the magnitude of the difference between randomly selected natural states of the system.[45]

The first concerns the basic physics of mesoscale and microscale phenomena. There are severe mathematical and theoretical difficulties in developing general prediction models for these smaller scales. The time scales are, of course, so short that the quasi-geostrophic relation between wind velocity and pressure is not valid. Furthermore, many of the weather systems are fully three dimensional, so that the hydrostatic approximation which is essential to general circulation theory cannot be applied to the smaller scales. Gravity waves are ubiquitous, and their effects must be accurately represented. Change of phase in clouds must be included. Also, the boundary layer plays a crucial role in many weather phenomena, so that the effects of turbulence must be incorporated in successful models.

The second difficulty concerns the observations required to describe mesoscale and microscale phenomena. In many cases, differences are small and the required accuracy of the observations is correspondingly high. Also, the data required to describe these small-scale features are numerous, and the time available for processing, interpretation, and dissemination of results is short.

In the face of all these difficulties, can one realistically hope to improve short-range prediction? Indeed, there are several avenues that should lead to major improvements. For periods extending from a few minutes to perhaps 2 h some mesoscale features can be successfully predicted by simple extrapolation of present conditions and rates of change. In order for extrapolation to be useful, the characteristic time scale for significant changes of the phenomenon must be larger than the time period of extrapolation. Thus, an individual wind gust whose time scale is a few minutes could not be reliably predicted for longer periods by simple extrapolation.

The crucial need is to apply modern technology to describe the local weather as it occurs and to communicate the desired information immediately to users. For example, by means of existing technology the motion of a squall line or the development of a thunderstorm could be displayed on any television screen, so that the viewer would be able to anticipate just what to expect at his home or along his route of travel. Power companies might be better able to anticipate sudden changes in customer demand given reliable information on significant short-period weather changes. At the present time, critical minutes or even hours often elapse between the time of observation and dissemination to the user. High-resolution satellite pictures often cannot be transmitted from data centers to users until days have passed. An important secondary benefit of a proposed

Local Weather Watch would be to provide detailed quantitative data for use in research on mesoscale and microscale phenomena.

At the upper end of the short-range prediction scale as shown in Figure 1 (8 to 12 h), improvements can be made by adapting present numerical models to the mesoscale region. This has been done to some extent using models that admit internal gravity waves. Such models employing small grid intervals have successfully represented development of fronts and other areas of concentrated baroclinicity.[46] Other phenomena of the mesoscale and microscale, e.g., vertical convection, clear-air turbulence, and secondary features of the planetary boundary layer, have not been incorporated into the numerical models. The physics of these weather features, especially their interaction with phenomena of other scales, remains largely unknown, and promising ideas are needed.

A third avenue along which increased effort should be channeled is the detailed analysis of mesoscale phenomena in their relation both to smaller-scale features and to the synoptic features in which they are imbedded. Technology provides measurement capability which has not been adequately employed on this problem. Cumulus convection, the structure of fronts or jet streams, clear-air turbulence, and the influence of mountain waves are among the important phenomena that can be examined with great precision using observations from satellites and aircraft and from radar and other ground-based remote-sensing systems. Greater use of this observing capability by well-trained research analysts is called for in order to describe accurately the phenomena to be predicted and their interactions with both larger and smaller scales.

SEASONAL AND CLIMATE CHANGES

We regularly witness marked changes in seasonal and annual weather. However, whereas we have a fairly complete theoretical understanding of phenomena on the synoptic and planetary scales as is indicated in Figure 1, understanding and the corresponding predictive capability are virtually lacking for variations of the seasonal and longer scales. We do not know, for example, the degree to which long-term variations may be inherent in an atmosphere with fixed boundary conditions and composition. Hypotheses exist that relate variations of the seasonal-to-yearly scale to interaction of the atmosphere with the upper layer of the ocean or with polar sea ice.[47] But adequate theories have not been developed, and numerical models have not simulated these variations, although encouraging beginnings have been made.[48]

Historical and paleobotanical evidence demonstrates that significant fluctuations in climate have occurred over at least the past ten thousand years.[49] We are not lacking in plausible hypotheses that may account for fluctuations in climate. For example, we can reasonably suppose that past changes in surface albedo accompanying the expansion of agriculture may have influenced climate to an important degree, but we do not understand the mechanism. Other hypotheses of climate change include the interactive coupling of the atmosphere with the deep ocean, changes in solar radiation received by the earth due to astronomical periodicities in the earth's orbit about the sun, and effects of volcanic eruptions on turbidity of the atmosphere. Despite the great importance of being able to predict accurately for these longer periods, we do not know whether any of these hypotheses is important in accounting for changes in climate.

General circulation models, upon being run for long periods, have successfully simulated many features of present-day climate.[50,51] This gives a degree of confidence in the realism of the models and suggests that they might be used to identify the principal causes of climatic change. However, physical factors that must have an important bearing on the problem have not been introduced. The sea-surface temperature is held constant instead of being allowed to respond to changes in the coupled atmosphere–ocean system. Although precipitation is computed, changes in cloudiness are not accounted for in the model; and the effects of clouds and aerosols on shortwave and long-wave radiation are ignored. External changes in solar radiation are also ignored. Models are being developed and tested which represent, at least selectively and crudely, some of the physical factors referred to here.[52] A few experiments have been conducted using arbitrary changes in CO_2 concentration; these constitute the first early steps in the prediction of possible future climate. As computing capability increases and understanding of atmospheric processes advances, it may become possible to include all factors expected to have a significant influence on climate and then to eliminate individual features in successive numerical experiments. As Lorenz points out, this should result in the identification of features or combinations of features that *could* have produced climatic changes. "As to what features *did* produce climatic changes, we shall still have the privilege of arguing."[19]

Simulation of climate using general circulation models requires that the full course of synoptic-scale weather systems be recreated in a computer for periods of at least several years. Climatological statistics are then compiled from the sum of all these weather states, which need not be individually accurate. This method is expensive in terms of computer time, and it is inconceivable that it could be used to study very slow fluctuations in climate. Furthermore, this brute-force method seems somehow inconsistent with the scientist's faith in simplicity; one would like to think that the procedure of experimental calculation should be no more complex than the physical hypothesis being tested. Lorenz has suggested that a hypothesis of climate change might be testable using a model with only a few hundred grid points rather than the 50,000 or more used in the most sophisticated models, even though the model might be quite unrealistic in features unrelated to the hypothesis being tested. Another approach might be to develop equations that result in filtering the individual synoptic weather features while accounting for their effects on the slower changes in climatological statistics. But neither of these possible approaches has been demonstrated.

The level of effort being directed at understanding our present climate and at anticipating possible future changes in climate is inadequate to the importance of the subject. The number of well-trained, able scientists working on problems of climate is quite small, and special efforts are justified to attract outstanding scientists and students to the field.

RECOMMENDATIONS

In assessing possible applications of the atmospheric sciences to human needs a number of factors must be considered in concert: scientific or technological feasibility, importance to public safety or welfare, urgency, economic or social value, probability of success, availability of necessary resources, and cost. There exists, of course, no formula yielding a priority index as a function of these factors, but certain general qualitative relations can be recognized. Feasibility must receive heavy weighting since it is essential to ultimate success, and necessary resources must be attainable. Where personal safety and welfare are relevant they should receive heavy emphasis, whereas in other cases considerations of economic or social value may be dominant. Urgency, probability of success, and cost are modulating factors which in many cases will be decisive. Minimum requirements for a responsible program include positive assessments of feasibility, probability of success, and availability of resources and either (a) importance to safety or welfare or (b) substantial economic or social value.

The Committee concludes that one of the major objectives of atmospheric-sciences research in the 1970's must be

To extend the capability for useful prediction of the weather and atmospheric processes.

This objective is placed first in this report because extension of predictive capability is necessary to provide the base that supports much of the fundamental research and the applications of the atmospheric sciences.

The recommendations that follow are addressed to the goal of improving prediction to serve recognized human needs. All scales have been considered. The order in which recommendations are listed represents our assessment of relative importance, taking into account the criteria indicated above.

Recommendation I-1 In order to extend the range of useful prediction into the one- to two-week period, *the international Global Atmospheric Research Program (GARP) should be vigorously supported and adequate resources should be provided promptly for necessary GARP experiments.*

Scientific plans for the U.S. contribution to GARP were outlined in the report of the U.S. Committee for GARP referred to earlier. That report emphasized the following requirements: (a) the development of a global observing capability; (b) the availability of electronic computers with speeds at least 100 times faster than the speeds of the most powerful computers in use today; and (c) the conduct of regional field programs and computer modeling experiments to improve the physical and mathematical basis of long-range prediction. The report developed these requirements in quantitative detail; it is therefore unnecessary to review that material here.

In most important respects, GARP activities are moving forward according to plan under the U.S. Committee for GARP of the National Academy of Sciences and the Office of Environmental Systems of the National Oceanic and Atmospheric Administration. There are, however, several specific problems that have been identified by USC-GARP and that require prompt attention in order that the GARP can be successfully carried out. These are

(a) Plans for data management and processing for the GARP experiments should be promptly developed.

(b) A strong numerical modeling effort is needed to carry out simulation experiments aimed at design of the global observing system.

(c) Economical means are needed to measure midtropospheric wind velocity in the tropics at one or more levels. The methods that have been proposed are expensive and technically uncertain.

(d) The following satellite instrument development programs, now under way, should be pressed strongly: temperature soundings below clouds and water vapor soundings (the latter is especially important in the tropics).

(e) Feasibility studies should be carried out promptly to determine the best method for making global measurements of wind velocity near the sea surface, soil moisture, extent of ice cover, extent of Arctic open water, precipitation, and water content of clouds.

The highest priority given to this recommendation reflects the high level of scientific and technical feasibility, and the importance of extended and more accurate forecasts to safety and to the economy. Necessary facilities and resources, though not all available, are accessible within 5 to 10 years, and the costs appear to be reasonable. Improvements in prediction models and the associated increased understanding of atmospheric processes will also contribute to the necessary basis for virtually every other application considered in this report.

A few study participants have dissented from the high priority assigned to the GARP program. They believe that the GARP is addressed to essentially scientific objectives, rather than to applications, or that greater benefits might be realized from direct efforts to improve short-range prediction.

Recommendation I-2 In order to determine the potential value of short-period extrapolations of local weather, *a pilot Local Weather Watch should be conducted for a selected urban area and a selected rural area that would provide real-time visual presentations of critical weather features.*

High-resolution satellite cloud images, ground-based weather radar pictures, and automatic station observations would be processed and coordinated by computer, and outputs would be disseminated and made available to any television screen in the region. For example, the approach of storms could be displayed graphically and accurately, air-pollution levels could be monitored, development of a sea breeze or an approaching sheet of cloud could be observed, or the detailed distribution of temperature or wind could be displayed. Provision would be made for specialized radio or television presentations for transportation, agriculture, and other businesses and industries, as well as for timely warnings of imminent weather dangers. The test would be run for a specific period, perhaps 3 years. The data collected in this program would provide a rich resource for research on a wide variety of problems concerned with the mesoscales and microscales.

A critical element in the program will be the provision of high-resolution satellite data in essentially real time. This will require a higher priority for these data than now exists for weather information. The system should be designed so that the users can interact effectively with the observing and dissemination system. To achieve this, a thorough, professional study should be carried out of the usefulness of the program to each major user group, its economic benefits and costs, and the degree of public interest and confidence in the product. The study should be an integral part of the program. It should be begun prior to initiation of the weather display service in order that results at every stage might be effectively incorporated in the design of the display and dissemination system and its operation. The study should consider the educational impact on the various user groups. Evaluation at the end of the pilot project should consider possible extension of the services to other areas and should include assessment of alternate modes of operational management, e.g., private industry, government licensed corporation, or government agency. This recommendation is addressed to the National Oceanic and Atmospheric Agency (NOAA), which is the agency responsible for the National Weather Service. NOAA should seek to enlist the cooperation of other federal agencies, local governments, industries, utilities, and radio and television stations in planning and operating the pilot Local Weather Watch.

Recommendation I-2 is assigned high priority in recognition of its technical feasibility, its potential contribution to safety from severe storms, and its probable economic value. It is regarded as an experiment that could result in a major improvement in weather services.

Recommendation I-3 In order to assess the possibilities of anticipating changes in climate and of making useful seasonal predictions, *increased efforts should be directed toward understanding the dynamics of climate.*

The largest available computer capacity should be provided for this purpose to a competent numerical modeling group as soon as possible. The numerical modeling efforts should consider all plausible hypotheses, e.g., ocean–atmosphere coupling, volcanic influences on radiation, astronomical effects on solar radiation. Effects of ocean–atmosphere coupling should be studied on the seasonal time scale as well as the climate scale. Special efforts are justified to provide adequate manpower with the necessary training to develop the field of climatic modeling.

Careful attention should be directed at monitoring the climate and at early and reliable detection of changes in climate. The effect of changes in trace gas concentrations and aerosols on the radiation balance of the atmosphere deserves special attention in view of man's ability to modify them. Long, homogenous records of high accuracy and reliability, at a few stations covering the entire globe, are indispensable. (See Chapters 3 and 4.)

This recommendation is important because the consequences of man's activities on climate must be known before the consequences become irreversible, and also because of the great economic importance of being able to make useful seasonal predictions. Developments in these fields are in their early stages, but there are reasons to believe that further useful results can be achieved in the next decade. It is especially interesting that, although only a few atmospheric scientists are working on the theory of climate, there is broad agreement that special efforts are justified in this area.

Recommendation I-4 In order to advance understanding of mesoscale phenomena and to improve ability to forecast these phenomena, *emphasis should be placed on research on fronts, jet streams, organized convection, and other phenomena of mesoscale.*

Increased support should be given to the development of numerical models capable of predicting such features as the weather associated with fronts and organized convection. The largest available computer in the hands of a competent numerical modeling group is needed for this purpose.

Strong support should be given to the development and testing of electromagnetic and acoustic remote-sensing techniques for measuring detailed distributions of wind velocity, humidity, temperature, and aerosol concentration. The International Field Year of the Great Lakes provides an excellent opportunity to test these techniques and to compare independent methods of measurement and should be used for this purpose.

Aircraft instrumentation should be substantially improved. Several research aircraft (about four) should be provided for mesoscale research with capability to measure fluctuations accurately in the three components of velocity, as well as temperature, humidity, drop size distribution, and radiation. These airplanes should be dedicated to research support and should not be responsible for operational requirements, such as hurricane reconnaissance.

A versatile, fast-response radiosonde (probably on a modular basis) should be developed for use in mesoscale investigations. The present National Oceanic and Atmospheric Administration's Boundary Layer Instrument Package (BLIP) or the Boundary Profile (BP) system of the National Center for Atmospheric Research should also be further developed as a modular instrument capable of making a wide range of measurements.

An economical, automatic ground station should be developed for making routine surface observations.

An area test should be conducted with the objectives of developing and testing mesoscale prediction models. Substantial economies would result from coordinating this test with related programs discussed in Chapters 2 and 3.

This recommendation is regarded as important in order to keep short-range prediction capability in step with capabilities at other scales. The complex physics associated with this scale of phenomena is formidable, so that this research should be regarded as a long-term effort which would benefit from increased but discriminating support.

Considerable discussion has gone into the ranking of this recommendation relative to Recommendations I-2 and I-3. The participants agree that mesoscale research deserves emphasis; however, specialists in mesoscale research and short-range forecasting, on the whole, believe that the recommendation should be assigned a higher priority. In the opinion of the Committee on Atmospheric Sciences, the complexity of the physics and of observational requirements justifies its present ranking.

A CHALLENGE FOR THE 1970'S

The last decade has been a period of unprecedented advances in the development of the technology and the science of prediction. Some of the successes have been dramatic, most have been the result of hard work and persistent efforts by large numbers of scientists and engineers. The scientific capability to extend the range of prediction to periods of a week or more, to provide much improved information of local weather changes, to examine systematically the mechanisms and causes of climate change, and to increase our understanding of phenomena on the scale of fronts, jet streams, and cloud clusters and how they interact with larger- and smaller-scale phenomena are all within reach.

There is every reason for confidence that well-planned, adequately supported research programs addressed to these goals should provide results of immediate benefit to public safety, the economy, and the management of environmental problems. It is clear that too little is known about the specific value of existing forecast services or of the added contribution that the programs recommended in this chapter can make to human needs.

These are among the outstanding challenges of the 1970's to the atmospheric sciences.

3

AIR QUALITY

Clean the air. Clean the sky. Wash the wind.

T. S. ELIOT

Among the problems of environmental utilization and management that have burst upon the public conscience in the past decade, probably none is more pervasive and potentially far reaching than the problem of air quality. Only the normal senses of smell, taste, and sight are required to detect the increasing frequency of haze-shrouded, odorous surroundings that are the product of our transportation systems, power generation stations, industries, and individual homes and apartments. We are rapidly learning about myriad combinations of toxic and deleterious materials, some of which react to produce still other pollutants. Many of these materials may change the ecological and climatological balance to which the population of the earth has attuned itself in the long course of its development. Air pollution has killed; it has produced among the susceptible, particularly the very young and the very old, chronic disabilities; and it has exacted material prices that give us reason to pause and reconsider our use of the atmosphere.

The more spectacular instances of fatal concentrations of air pollutants have been well documented and well reported. The London disasters of 1952 *et seq.*,[53] the Donora incident,[53] and the New York City incident during the Thanksgiving of 1966[54] are instances in which air pollution was responsible for loss of human lives. These episodes occurred in limited regions where the sources of the pollutants and the victims were located together and the pollutants accumulated to unusually high concentrations leading to the resulting illnesses and deaths. These high concentrations of toxic materials were directly attributable to unusually low atmospheric dispersion.

Health effects short of fatality but sufficient to produce clinical evidence of increased morbidity were also clearly associated with these air-pollution incidents. Cardiorespiratory difficulties were particularly evident, and in most cases several days and even weeks, depending upon the predisposition of the victim to these difficulties, were required for the victim's recovery after the pollution was reduced to more tolerable levels.

Effects of air pollution on health can also occur over wide areas at lesser but persistent levels of pollutant concentration and are gradual rather than sudden and dramatic in their impact. They constitute another class of air-pollution problems beyond the air-pollution episode.

In assessing the biological effects of pollutants attention usually focuses on airborne materials that emanate from man-made sources. There are also a number of "natural" pollutants that produce significant biological effects. Prominent among these are the aeroallergens and spores and fungi that transmit plant diseases. In many cases these pollutants are responsible for highly adverse effects on human comfort and biological productivity. Other sources of significant pollution are forest fires, volcanic eruptions, and wind erosion of deserts and cultivated fields.

In addition to the direct and indirect effects of air pollutants on human health there is a long catalog of effects on animate and inanimate objects of value to man. A recently published pictorial representation of known effects of pollutants on plant life[55] provides an excellent measure of the diversity and severity of these effects.

Recent and still incomplete studies have shown that pesticides may be airborne far beyond their point of application. By the processes of biological magnification, i.e., by successive biological concentration in food chains and webs, some of these materials are threatening the reproductive success of birds and are accumulating in animals to the point where their fat and milk products are unfit for human consumption. From these and other examples we have begun to identify problems of ecological balance and direct threats to human food supplies that are related to airborne transport, dispersion, and deposition of man-made pollutants.[56]

Inanimate objects suffer from exposure to air pollutants in many ways. Fabrics are weakened, paint is pitted and discolored, stone and metal are softened and corroded, glass is etched, and rubber is cracked. Accumulations of particulate matter and some gases produce a formless haze that reduces visibility, thereby hampering transportation systems and denying the satisfaction of distant vistas and clear sunshiny days. Attempts have been made to assign a price tag to some of these effects, particularly the costs incurred for the maintenance of property and the level of cleanliness demanded by most American families.[57] Because of the diversity of these costs and choices that individuals make in allocating finite resources, no "hard" estimate is possible. But recent estimates place these material costs between $10 billion and $20 billion per year in the United States. The allocation of these costs is, of course, directly related to the local concentrations of pollutants, and they therefore constitute a selective economic impact on urban and industrialized areas.

As noted above, the most dramatic evidences of polluted atmospheres are the sporadic episodes of locally high concentrations of multiple, man-made pollutants. These episodes are scattered among more frequent days of tolerable, often unnoticed, lesser concentrations of these materials. Normal variations of atmospheric ventilation and mixing do, indeed, provide a range of local accumulations of pollutants, even with constant sources of these pollutants. However, beyond local accumulations we must also be concerned with the accumulation of pollutants in the global atmosphere. These accumulations could conceivably bring about subtle but far-reaching changes in global climatic and ecologic regimes or even set in motion irreversible changes that could lead to catastrophe.

Management and control of air quality requires scientific and technological contributions from chemists, biologists, medical doctors, and engineers, as well as from atmospheric scientists. Ultimately, however, air quality is a public policy problem. The public must decide, through broad discussion and the normal channels of decision making, what forms of air pollution man should be expected to tolerate and how much he is willing to pay for smog-free cities.

Air quality poses a hierarchy of problems that range from one individual's comfort to the ultimate welfare of the entire world's population. In this global context, the problems of air quality transcend the ethnic, cultural, and national interests and become, quite literally, a "common threat" to all mankind.

THE ROLE OF THE ATMOSPHERIC SCIENCES IN THE AIR-QUALITY PROBLEM

In addressing the multiple problems of air quality, atmospheric scientists have a critical and central, but not an exclusive, role to play. Boiled down to bare essentials, regional, national, and international air-resource-management programs must operate within the finite capacity of the atmosphere to disperse, store, and deposit waste materials and their residual reaction products. Two variables may be controllable in this management scheme: (1) the rate of injection of each pollutant and (2) the place of injection. The management scheme must optimize the rates and places of pollutant discharges by balancing the benefits of utilization of the atmosphere for waste disposal against the costs of adverse effects of primary pollutants and their reaction products. The evaluation of these benefits and costs requires consideration of social, economic, and political, as well as scientific, factors. Eventually, however, there must be an allocation of the right to use the atmosphere for waste disposal which reflects considered assessment of alternative costs and benefits and deliberate judgments regarding acceptable risks.

The multiple problems of air quality also require unified consideration of pollutant effects on several scales of time and distance. A plan that controls catastrophic consequences of local air-pollution episodes does not necessarily control chronic regional effects and *vice versa*. Plans that control chronic regional effects may not be sufficient to control the global dispersion and accumulation of pollutants. And, finally, control procedures appropriate for one pollutant may be totally inadequate for another because of their differing reactivities, residence times, and effects on animate and inanimate objects or on climate.

The central role of the atmospheric sciences in this complicated problem is to specify, on all useful scales of time and space, the capacity of the atmosphere to absorb, disperse, modify, and deposit pollutant materials, given any arbitrary arrangement of geographic positions and rates of discharge of multiple pollutants. When compared with the effects and acceptable levels of risk, these determinations provide the basis for choosing acceptable modes of discharge of pollutants to the atmosphere, and therefore they enter as significant criteria in urban, regional, and general land-use planning and development.

The primary questions that the atmospheric scientists must address in making these determinations are the following:

1. What materials are to be emitted to the atmosphere? What is their chemical composition and physical state, i.e., gas or aerosol? What are the size distributions of the aerosols?

2. Where and at what rates are these materials emitted? At what temperature and height above the surface is the effluent discharged?

3. How are these materials transformed, chemically and physically, during their residence in the atmosphere?

4. Where are the primary and secondary pollutants transported? What are their concentrations?

5. Where, at what rate, and by what processes are pollutants removed from the atmosphere?

6. Do the predicted concentrations and deposition accumulations exceed acceptable values? If so, what changes of source locations and strengths will correct the situation?

In addressing these air-quality problems, the atmospheric scientist must also consider the possibility that the presence of pollutants, particularly sensible heat, may alter the processes of pollutant transport, diffusion, and chemical and physical reactions. Viewed in their entirety, these problems involve complex, nonlinear systems which can be adequately treated only by sophisticated computation and measurement schemes.

Because airborne pollutants may remain airborne for extended periods and travel over great distances, it is convenient to discuss the problems they create as *urban, regional,* and *global. Urban* problems are associated with the local accumulations of pollutants caused by single or multiple sources located in urbanized or industrialized areas. Maximum concentrations and dosages are generally associated with these largely localized systems. *Regional* problems are associated with the accumulations of pollutants as the air in which they are embedded moves over sequential source and nonsource areas. Retention of pollutants as the air moves from source area to source area aggravates the urban problem, but these residual pollutants also produce effects in the nonsource areas. *Global* problems are associated with residual accumulations of pollutants in the global atmosphere and are an extension of the regional problems. Global problems are generally concerned with climatic and biological or ecological effects of worldwide alteration of atmospheric constituents. These distinctions are, of course, somewhat arbitrary—the atmosphere and its pollutant load represent a continuum of properties and problems which interact on all scales.

The Urban Problem

The basis of a rational control policy for urban air quality is a determination of the contribution of each source of pollution to the concentration of pollutants in the air of cities and their surroundings. This must be known for both long-term averages and short-term extremes. The problem is attacked by constructing urban and regional air-quality simulation models. A basic input for these models is an inventory of either actual or proposed emissions. Given the sources, the concentration of emitted pollution from each source is traced as it is transported by the larger-scale wind motions and dispersed by the organized mesoscale motions, mixed laterally and vertically by the smaller-scale turbulent motions, and as it is removed or transformed by fallout, rainfall, or chemical reactions. There is already sufficient experience with these models to show their great potential utility. Their further improvement depends upon acquiring, through research, more precise knowledge of the following critical features of the air-pollution system[58]:

1. Chemical and physical reactions of pollutants while airborne;
2. Methods, places, and rates of removal of pollutants by wet and dry fallout and deposition; and
3. Variations of atmospheric transport and mixing with height over over urban areas.

There is a good basis for the belief that models will eventually be produced for the control of air quality during high-concentration emergency, as well as others that will be essential tools in rational urban and regional planning.

Highest priority should be given to the development of simulation models and to the supporting experimental work that makes the models reliable tools in air-quality analyses. The most acute problems from the point of view of the direct effects of air pollutants on human welfare, are clearly associated with accumulations of pollutants within contiguous urban areas. Most of the present efforts of regulatory agencies are directed toward abatement of these more severe problems.

Among the multiple atmospheric problems associated with urban air pollution, the role of chemical reactions among airborne pollutants has emerged as critical to successful understanding and simulation of air-quality systems. This field has been neglected by atmospheric scientists, despite the fact that atmospheric transport and mixing are intimately involved in these reactions and cannot, themselves, be properly assessed unless the reactions are known. An example of the current degree of ignorance of urban atmospheric chemistry is that little is known about what happens to ubiquitous pollutants such as SO_2 and CO in the presence of other pollutants in the open atmosphere. Similarly, attempts to simulate the interaction of atmospheric mixing, chemical reactions, and solar radiation which produces photochemical smog are in a most primitive and unsatisfactory state. As long as this is the case, the ability to anticipate the air-quality consequences of proposed abatements and relocations of pollutant sources will be unsatisfactory.

In order to establish the reliability of simulation models of urban air-pollution systems, including chemical reactions and deposition processes, a comprehensive and well-coordinated experimental program which accounts for real pollutants under natural conditions is essential. Present methods for field measurement must be supplemented by applying remote-sensing technology to the measurement of air pollution and the associated atmospheric properties. Astronomical observations have been used to demonstrate that changes in urban pollution concentrations can be measured remotely.[59] Absorption spectroscopy using ground-based and satellite instrumentation offers especially promising possibilities for measuring pollutant concentrations and distributions. Measurements on some typical pollutants indicate that their concentrations can be determined to within a few parts per billion.[60] The development of these techniques, which are badly needed on all scales, should be vigorously pursued. A comprehensive field measurement program is needed to compare *in situ* and remote-sensing observations and to validate the analysis systems necessary for air-quality control and planning.

The Regional Problem

The primary air-quality problems that are properly assigned to regional considerations are due to airborne materials accumulated in the sequential passage of air over one or more pollutant emission areas, generally involving transport over distances of hundreds of miles. In the limit, these accumulations become global, of course, but prior to that stage they pose air-quality problems that in some cases are international in scope. Improved understanding and prediction of regional air quality depends, in part, upon better definition of atmospheric transport and dispersion and on the fate of pollutants over these travel times and distances. The rates at which pollutants react and are deposited are critically important.

No systematic investigation of the spread of air pollutants on scales of a few hundred miles has yet been made, but work has begun. Some current numerical weather-forecasting methods lend themselves to an attack on this problem by means closely analogous to the urban air-pollution model. The shortcomings of this method are analogous to those of the urban models. In the first place, the transformations of the pollutants in the atmosphere must be known. Of particular interest is the conversion of SO_2 to H_2SO_4 and sulfates in the presence of other contaminants. In the second place, details of the wind-field and mixing processes on the mesoscale may be important in the determination of the paths of pollutants. Current trajectory techniques tend to smooth out important variations of pollutant travel. Solution of this transport problem is required for rational planning of the distribution of future industrial complexes associated with population growth.

The regional problem, as defined in this report, is less pressing than the urban problem and is in an earlier state of specification. Because regional problems focus on larger-area effects and involve longer time averages of pollutant concentrations, they require a different scale of measurements from those required for localized urban problems. However, investigation of the transport and fate of selected pollutants could be coordinated with the comprehensive urban measurements program discussed in the preceding section.

The Global Problem

Global air-pollution problems result from the contributions of many natural and man-made sources of significant pollutants. The sources include fossil fuel combustion all over the world, isolated volcanic eruptions, and detonations of nuclear devices. In addressing problems associated with the accumulation of pollutants on the global scale, the research needs already discussed within the urban and regional contexts must again be considered. In addition to the impact that significant changes in weather and climate may have on man and the biosphere (see Chapters 2 and 4), the global spread of toxins also poses a significant stress on the ecological balances among living organisms. For example, the spread of DDT to all parts of the globe has occurred so quickly that it must have been transported via the atmosphere. This is only one example. Other toxins, such as heavy metals and trace gases, also enter into considerations of global bio-environmental problems. The useful prediction of the distribution of these materials as their sources proliferate is an essential tool for international management of atmospheric pollution problems.

The most pressing need on this scale is the determination of significant pollution concentrations by a global measurements system. Not enough is known about the levels and variations of these concentrations to make it possible to assess the global impact of man's activities. Baseline measurements of pollutant levels as they relate to ecological systems are essential. Over a period of time such measurements should portray natural variations in these systems as well and make possible an assessment of man's contribution to such changes. If we are to develop this level of understanding and the capability to predict the consequences of further human intervention, it is imperative that the atmospheric processes involved in distributing pollutants to the biosphere also be adequately measured. The result should be a reliable global atmospheric dispersion model, an essential ingredient of any comprehensive global ecological simulation model.

RECOMMENDATIONS

One of the major objectives of the next decade should be

To contribute to the development of the capability to manage and control the concentrations of air pollutants.

This objective is given priority just below extension of weather prediction because of the urgency of effective and equitable management of waste disposal to the atmosphere in order to avoid harmful effects on human health, ecological systems, and potential climatological changes.

In order to achieve this objective the following actions are recommended in order of their importance:

Recommendation II-1 In order to develop reliable methods to assist in the assessment of alternative urban pollution control programs, *(a) development and testing of simulation models of the urban source–atmosphere system should be accelerated,* and *(b) a major field program of at least two years' duration, designed to measure in detail the fate of pollutants emitted from an urban area, should be carried out.*

An excellent start has been made in the development of urban air-pollution models. An accelerated effort is feasible and, in view of the urgency of these problems, required. The field measurements program recommended, in addition to emphasizing the processes of chemical reaction, phase change, and deposition of natural pollutants, must also provide clear definitions of the processes of atmospheric transport and diffusion. These measurement programs are intended to provide data necessary to validate and improve urban simulation models. In view of the variety of pollutants, terrain, and weather regimes that must be considered, detailed studies of the fate of specific pollutants should be carried out at several sites.

The accomplishment of the measurement program depends heavily upon development of remote-sensing systems for the measurements of boundary-layer winds, vertical temperature profiles, and pollutant concentrations. We, therefore, further recommend that, in addition to significantly increased support for the National Oceanic and Atmospheric Administration (NOAA) Wave Propagation Laboratory and for the development of satellite instrumentation by NASA, support should be provided for three or four university groups devoted to the development of remote-sensing techniques and to their application in boundary-layer research. In order that proven systems be available for the field programs during the mid-decade, NOAA's Wave Propagation Laboratory effort should be accelerated immediately.

Recommendation II-2 In order to assess the potential effects of the global distribution of pollutants, *an effective system for global monitoring of atmospheric contaminants should be established promptly.*

This program must be coordinated with the total program of global environmental monitoring which is under active consideration by a variety of scientific disciplines. Atmospheric scientists must be responsible for the design of the atmospheric measurements.

Recommendation II-3 In order to understand the processes that determine air-pollution concentrations on a regional scale, *the chemical composition of precipitation and the dry fallout should be measured for one or more selected areas and periods.*

In recommending this program, a network of 50 to 100 air and precipitation sampling stations is envisaged arranged over an area that extends 100 to 150 miles from the primary source areas. Centralized chemical analyses facilities, patterned after the present Air Pollution Control Office (APCO) national facility, should be employed. Pollutants to be measured should be chosen on the basis of residual effects both on the regional scale and on their potential for accumulation in the global atmosphere. SO_2 and its reaction products, heavy metal, suspended particulates, and pesticides are prime candidates.

These measurements, while considerably less detailed than those required for urban pollution studies, should be compatible with the intensive studies of airborne chemical reactions and deposition recommended for the urban problem (Recommendation II-1) and may be advantageously made concurrently with that program. Since both of these programs are aimed directly at quantitative assessment of the "self-cleansing" properties of the atmosphere, they are critically important to the managerial problem of allocating the use of the atmosphere for waste disposal on an equitable basis.

FURTHER IMPLICATIONS FOR RESEARCH

Beyond these explicit recommendations, there are two facets of the activities within the atmospheric sciences associated with air-quality problems that deserve national attention. The first, which derives directly from the interdisciplinary nature of these problems, is a clear need for systematic evaluation of the costs and benefits of atmospheric waste disposal, and out of these evaluations, an authoritative assignment of urgency among these numerous pollution problems. Or in the phrases of this study, what are the "human needs" for air quality that the atmospheric sciences, in league with other disciplines, should address? We have probed far enough into these requirements to know that a substantial amount of research is required before the basic human value judgments, attitudes, and priority assessments that play a large part in this problem can be adequately specified. We urge strong support for such research with the objective of setting authoritative guidelines for priorities in both air-quality research and in the application of the results to air-quality control programs.

Second, we have identified the critically important role of atmospheric chemistry in air-quality problems. To make effective progress in this area, it is most important that a few universities develop significant atmospheric chemistry research and graduate education programs. We emphasize the association of the field of chemistry with atmospheric processes and the need for joint consideration of the dynamics and physics of these processes with the chemical processes involved. This merging of knowledge and inquiry among what have heretofore been identified as separate, specialized fields reflects the fact that important problems have now reached the point of complexity where only concerted attacks can successfully cope with them.

4 WEATHER AND CLIMATE MODIFICATION

We dare not proceed blindly—the risk of obtaining insignificant results is too great.

HENRY STOMMEL

Some 25 years ago, a series of field experiments initiated by scientists of the General Electric Company demonstrated that deliberate modification of clouds and precipitation was possible. Since that time, weather modification has received a great deal of attention by atmospheric scientists and by the public. Laboratory and field investigations have expanded understanding of the physics of clouds and precipitation. The business of operational cloud seeding has grown up, and extensive projects have been carried out in many countries for purposes of increasing or redistributing rainfall, for suppressing lightning and hail, and for dissipating fog. Weather modification has been a center of controversy during much of the past 25 years. Scientists have disagreed on the interpretation of experiments and operational projects, operators have made contradictory and sometimes extravagant claims, lawsuits have been filed to recover damages or to prevent cloud-seeding efforts, federal and state government policy has been uncoordinated, and, in part, contradictory, and understandably, the public has been confused by it all. In recent years inadvertent modification of weather and climate through human activities has provided a new topic of growing concern.

The technical reasons for the long-continued controversy over cloud seeding, despite some notable advances, are that (a) we do not have a complete and accurate understanding of how clouds create raindrops and snowflakes, (b) field observations usually have been inadequate to describe accurately the specific changes resulting from cloud-seeding efforts, and (c) the effects of individual seeding experiments, even when successful, usually fall well within the natural variability of clouds and cloud systems.

The progress made on the scientific and technical problems of weather modification has brought the subject to the point where certain significant effects can be identified and programs can be designed for examining systematically the potential value of large-scale weather modification efforts. However, as the technical problems are overcome, another group of problems looms larger. These are the problems of public policy associated with weather modification. Examples of these problems include the economic value of cloud seeding as opposed to alternate means of increasing water supply or agricultural production, determination of responsibility for damages from cloud seeding, definition of the roles and responsibilities of government and private industry, criteria for deciding what operations are in the public interest, ensuring that long-range aspects are fully considered, and providing a mechanism to cope with international ramifications. The science and technology of weather modification are responsible for determining what is possible, for suggesting effective means for modifying weather, and for anticipating possible deleterious effects of inadvertent modification before they occur. The crucial question of what weather modification efforts are in the public interest cannot be settled by atmospheric scientists alone. This question involves economic, legal, political, and administrative, as well as scientific and technical, considerations. Involved also is the whole complex of environmental issues to which the nation will have to respond in the coming decades: issues of land usage, of the degree of freedom to use and affect the atmosphere, procedures needed to reach equitable decisions in cases of conflicting interests, and methods needed to safeguard the future ecology and the climate.

A comprehensive review of the scientific and technical aspects of weather and climate modification was carried out five years ago by this Committee,[6] and a further review is now being completed.[41] The results of these reviews have been used as the primary scientific and technical inputs to this chapter. Because an up-to-date comprehensive review of the science and technology of weather modification will be published soon, it is unnecessary to discuss those aspects in detail. Therefore, recent achievements and remaining critical problems are briefly summarized in the next two sections. Emphasis is placed in this report on the administrative and social issues that will strongly influence public policy in the area of weather and climate modification.

ACHIEVEMENTS OF THE PAST DECADE

During the past decade, research effort in weather modification has pushed forward on many fronts. It has been a period of slow clarification of concepts and ideas rather than of dramatic new discoveries. The most important achievements appear to have been the following:

Modifying Precipitation

There has been a steady accumulation of statistical evidence supporting the conclusion of the 1966 report[6] that "there is increasing but still somewhat ambiguous statistical evidence that precipitation from some types of cloud and storm systems can be modestly increased or redistributed by seeding techniques." Randomized experiments have indicated that ice-nuclei seeding may in some situations lead to increases in precipitation amounting to 10–30 percent and in other situations to corresponding decreases in precipitation. In other instances, cloud seeding has had no discernible effect. The characteristics that distinguish increases from decreases for orographic clouds appear to be associated with the temperature and wind distributions, but the role of these distinguishing characteristics in influencing the effects of seeding has not been adequately explained. Distinguishing characteristics for convective clouds appear to be related, at least in part, to the stability characteristics of the cloud environment, but much more research is needed in this area. There is little evidence relating to the effects of seeding on precipitation in large-scale nonorographic cloud systems.

Analysis of Convective Clouds

The use of ground-based radar has made possible substantial advances in the quantitative analysis of the fields of motion, liquid water, and ice in convective clouds. Development of more sophisticated techniques using two Doppler radars has made possible still more detailed descriptions of the three-dimensional distribution of the wind field. Satellite pictures have revealed organization of convective clouds on a variety of scales.

These developments remain to be fully exploited. Direct measurements of storm environment together with the radar-derived wind field can provide improved knowledge of the modifications produced by cloud-seeding experiments. Further development of remote-sensing techniques using satellites and radar should make possible the examination of the dynamics of convective clouds and their interaction with their environment.

Modifying Dynamics of Convective Clouds

Numerical models of convective clouds have been developed that can be used to predict dynamic effects of the release of latent heat from ice-crystal growth initiated by seeding. Numerical experiments can be used to design seeding operations intended to modify the location and intensity of the updraft and the height of the top of the cloud. A limited number of field tests of such models have been carried out and have yielded positive results indicating that the vertical extent of some convective clouds can be increased by means of ice-nuclei seeding.

Modifying Dynamics of Hurricanes

Simple, circularly symmetrical numerical models of hurricanes have been used to simulate the development of hurricanes following specific modes of cloud seeding. These numerical experiments suggest that significant effects may result from seeding. A field test of Hurricane Debbie in August 1969 has indicated decreases of maximum wind speeds of 15–30 percent following silver iodide seeding.

These limited results suggest that it may be possible to reduce the damage inflicted by hurricanes and possibly to influence the path followed by these storms. Because of the great potential value of these efforts, it is urgent that more realistic hurricane models be developed and that further careful field tests of model predictions be carried out (see Chapter 5).

Fog Dissipation

Dissipation of supercooled fog by seeding with dry ice, silver iodide, or liquid propane has become an operational practice at several airports in the United States, the Soviet Union, and France. Success rates of 60–80 percent have been reported. However, because supercooled fogs represent only about 5 percent of the fogs occurring at airports in the United States (excluding Alaska), increasing efforts have been directed at dissipation of warm fogs. Some successes have been reported using a variety of means, e.g., seeding with very large hygroscopic nuclei (about 20 μm), downwash from helicopters, exhaust from jet engines; but a proven, practical method for operational dissipation of warm fogs does not yet exist.

Hail Suppression

There have been widespread efforts to mitigate damage caused by hailstorms. Large numbers of ice nuclei are added to the cloud in order to reduce the size of hailstones by increasing their number. Spectacular successes have been reported by investigators in the Soviet Union over a period of years in protecting crops from destruction by hail. In these experiments, seeding agents are introduced directly into the supercooled region of the clouds by rockets or artillery shells. The results of ice-nuclei seeding in other countries, including the United States, have been mixed. The National Hail Research Experiment has been organized and will begin field operations in 1972 under the management of the National Center for Atmospheric Research (NCAR). This program is intended to test, with carefully designed field experiments, the feasibility of reducing hail damage by seeding.

Inadvertent Modification

At the beginning of the 1960's interest in the inadvertent modification of weather and climate was limited for the most part to discussions of the effect of the increasing CO_2 concentration on temperature of the lower atmosphere. Conclusions were variable and of uncertain significance. During the ensuing decade, models have been developed that are more realistic than the earlier ones; recent calculations using these models suggest that the expected increase of perhaps 18 percent in atmospheric CO_2 by the year 2000 might result in a surface temperature increase of 0.5°C.[56] This is comparable to the natural variability of climate over 30 years; consequently, the CO_2 problem is not likely to be critical in the next few decades.

Over longer periods, in which CO_2 concentration may continue to increase, the climatic effects may be very important.

The effects of particulate matter on radiation have also received attention in recent years. There is evidence that over the past three decades atmospheric turbidity in the northern hemisphere has increased measurably, while direct solar radiation has decreased. Electrical-conductivity measurements indicate substantial increase in particle concentration over the North Atlantic in the past half century with, however, no corresponding increase in the South Central Pacific. The effects of these changes in particle concentration on global temperature may be substantial, but the sign of the temperature changes has not yet been determined. The results depend critically on refinement of knowledge of the optical effects of atmospheric particles. Increased particle concentrations may also result in increased cloudiness and additional radiative effects of possible importance. Effects of emissions from SST's flying in the stratosphere are receiving continuing attention.

Urban and regional weather and climate have experienced substantial changes as a result of increased air pollution. Cities typically experience measurably higher temperature and more clouds, fog, and precipitation than adjacent rural areas.

Other effects on the weather and climate of the future, which have received consideration, include changes in surface albedo, roughness, or conductivity; heat output of cities; and the effects of oil films on water surfaces. These are important locally, but on the global scale they appear to be less important than the effects of CO_2 and particulate matter. Conclusions, however, are still tentative.

MAJOR CURRENT PROBLEMS

Critical Technical Problems

Further progress in weather modification will require increased understanding of certain basic processes and improved capabilities for measurement. The following problems are recognized as especially critical:

Microphysical Processes The microphysical processes associated with the production of precipitation have proved to be quite resistant to efforts to understand them. Careful laboratory experiments have provided accurate data concerning these processes, but there remain fundamental aspects that are not understood. Understanding of the various modes of ice nucleation remains qualitative, and specific predictions of effects of nucleating agents have not been made. The processes by which ice crystals multiply in clouds are still obscure, as are the process or processes of electrical charge separation in clouds. It is essential to improve understanding of the interactions of microphysical and mesoscale features of clouds and cloud systems.

Cloud Measurements Reliable methods for measuring the microphysical properties of clouds, condensation-nuclei and ice-nuclei distributions, drop-size and ice-crystal distributions, liquid-water content, turbulence velocities, temperature, and electrical charge distribution are badly needed. *In situ* methods using aircraft exist for making some of these measurements, but they are not yet sufficiently accurate or reliable; they must be further developed and tested. At the same time, existing techniques using radar, lidar, and other remote sensors must be further developed in order to be able to make the necessary measurements in regions of severe turbulence where aircraft cannot fly and to obtain complete distributions over extensive volumes. *In situ* and remote measurements must be carefully and systematically compared in order to develop confidence in the results of measurement.

Further Clarification of Cloud-Seeding Results Before we can have confidence in planning extensive cloud-seeding operations, the results of cloud-seeding experiments must be understood more specifically. The physical basis for the separation of positive from negative changes in precipitation must be understood more completely and precisely for orographic, convective, and large-scale nonorographic cloud systems. This will require improved understanding of the microphysical processes in clouds, improved methods for cloud measurements, further development of cloud models, and additional carefully designed field experimentation.

Hurricane and Convective Storm Models In order to determine the possibilities of beneficial modification of severe storm dynamics, it is important that the simple models of hurricanes and convective storms that now exist be developed into more realistic models. These models should be capable of accurately predicting important properties of hurricanes and convective storms. Field experiments will be needed to test the accuracy and usefulness of such models and to contribute to systematic improvement in the models (see Chapter 2).

Observations of Gases and Particulates In order to know what effects on the global climate are being produced by changes in concentrations of various particulates and gases, it is essential that accurate global data be acquired at sites that are representative and likely to remain representative for extended periods. Observations should be standardized, and instruments should be calibrated on a regular schedule. The atmospheric quantities required for investigations of global climate should include the following: CO_2 concentration, aerosol concentration, electrical conductivity, and total solar radiation. These observations should be closely coordinated with the program of global observations proposed in Chapters 2 and 3.

Administrative Problems

At the beginning of the last decade, the National Science Foundation (NSF) was charged in Public Law 85-510 (1958) with responsibility to "initiate and support a program of study, research, and evaluation in the field of weather modification" and "to report annually to the President and the Congress thereon." In this way emphasis was placed on research into the scientific problems associated with weather modification, and NSF was assigned lead responsibility within the federal government for supporting and reviewing research in this field. Under this mandate, the funds allocated by NSF, chiefly to university and NCAR scientists, have increased from approximately $1.3 million per year at the beginning of the decade to $3.2 million in fiscal year 1970, a modest ten-year increase for a field receiving special emphasis. At the same time, considerably greater increases in funding have occurred in other federal agencies, notably the Departments of the Interior and Commerce and the Navy Department; the total annual budget supporting research in weather modification has increased from $2.4 million in fiscal year 1961 to about $12.0 million in fiscal year 1970, as listed by the Interdepartmental Committee for Atmospheric Sciences (ICAS). In addition to support for basic research, these figures include support for programs directed toward applications. The Department of the Interior's Bureau of Reclamation has supported work directed at increasing water supply in the arid west; the Department of Agriculture has supported work directed at lightning suppression; the Navy, the National Aeronautics and Space Administration, and the Air Force have supported work directed at fog dispersal; and the Navy and Commerce Departments have supported severe storm modification. Other agencies have supported similar applied objectives. Recognizing the broadened objectives of federal agency activities in weather modification, in 1968 Congress passed and the President signed Public Law 90-407, which encouraged efforts directed toward a variety of agency missions in addition to scientific research and which left to the Executive Branch the development of a coordinated program. There is now no designated lead agency for weather modification.

A policy of some diversity of support in a field like weather modification has real advantages. An agency that is responsible to the nation for providing water supplies for future development of cities or for irrigation or for protection from severe storms or from forest fires, for example, must be able to evaluate the potential contribution of weather modification to its mission. This requires close contact between the agency and research scientists, and it may require the agency to support or to carry out mission-oriented research projects. If research convinces the agency of the value of weather modification, it should be responsible for presenting its case for an operational program to the Executive Branch, the Congress, and the public. This is the policy, admittedly *ad hoc*, that has developed in weather modification.

But in order that the 11 separate agency programs constitute an effective national program, coordination is essential. Responsibility for interagency coordination of research was assigned in 1959 by Presidential Executive Order to the Federal Council on Science and Technology (FCST) and for weather modification to the Council's Interdepartmental Committee for Atmospheric Sciences (ICAS). This mechanism, together with annual interagency conferences on weather modification originally organized by NSF and now by ICAS, has provided regular exchange of information and discussion among agencies. Annual ICAS reports have included summaries of agency programs in weather modification, and FCST has reviewed these program budgets before final submission to the Bureau of the Budget (now the Office of Management and Budget).

However, FCST and ICAS have not been able to develop an integrated national program in weather modification. Individual agency programs have been subcritical in size and research capability. ICAS has no authority to consolidate or to modify agency programs; and, most important, ICAS is not able itself to mount research efforts, no matter how badly needed they may be. Agency initiatives at the scientist level, even though endorsed by ICAS, may not be approved by agency administrators; and agencies may launch major programs without ICAS endorsement. The result has been that in important respects the national effort in weather modification has been largely dissipated in submarginal projects, while crucial problems requiring large programs remain unsolved.

Despite widespread recognition of the need for stronger coordination in weather modification, this administrative problem remains as one of the crucial blocks to an effective national program. Creation of the National Oceanic and Atmospheric Administration (NOAA) will not necessarily help this problem, since the major weather-modification programs remain in separate agencies.

In considering how best to solve this problem, the following requirements are central:

(a) Responsibility for research in weather modification must be closely associated with responsibility for research in the atmospheric sciences generally. This follows from the indubitable fact that solution of critical scientific problems in weather modification requires sophisticated understanding of a wide range of atmospheric processes.

(b) Agencies may have a legitimate need to conduct specific mission-oriented research and must be responsible for the development and support of these programs.

(c) National policy in weather modification must be based on full consideration of relevant economic, social, ecological, and legal factors, as well as on scientific and technical factors.

A suitable administrative solution consistent with these requirements would be to assign lead responsibility for research in weather modification and for coordination of major field programs to a single agency and to encourage other agencies to continue to support programs within their specific areas of responsibility.

Public-Policy Issues

The public-policy questions associated with weather modification have been the subject of study by several groups which have included economists, ecologists, lawyers, statisticians, political scientists, geographers, and sociologists in addition to atmospheric scientists.[61–67] These studies have responded to such questions as the following: Is increased precipitation from extensive cloud seeding beneficial or harmful; under what circumstances and to whom? Should the value of increased water supply outweigh the general public's desire for dry roads and more sunshine? Should water supply, transportation, or ecological considerations get first priority? Who will soothe the angry losers or pay for damages? Each of these questions may have international as well as national, regional, and individual implications.

There are, of course, no definitive answers to these questions. Meaningful answers are possible, at the most elementary level, only after the scientists and technologists are able to reduce the uncertainties regarding the effects of specific weather-modification efforts. Therefore research on the physical problems must continue to receive first emphasis. Apart from this limitation, the questions posed above cannot be answered definitively because they are intrinsically involved with the conflicting interests of individuals and groups. Clearly, in order to ensure that weather modification is directed toward appropriate human needs, such problems will demand the highest level of intelligence and professional skill. The problems are interrelated with many subjective elements. They will have to be *coped with*, rather than *solved* in the manner familiar to physical scientists. The following brief statements attempt to identify the key elements of these problems as they are recognized at this time.

Economic Aspects There are numerous examples of projects for which attractive benefit-to-cost ratios are quite likely, assuming technical success of cloud seeding at the level now suggested by available evidence. These include augmenting water supplies for power generation and irrigation, reducing the occurrence of lightning-caused forest fires, increasing agricultural yield, and dispersing cold fog over airports and highways. Less certain technically, but potentially of the greatest economic value, would be the effect of reducing the wind speeds associated with hurricanes and other severe storms. Nevertheless, little effort has gone into economic evaluations of weather modification, especially in comparison with the large private expenditures that have been applied to operational cloud-seeding projects. The reasons are straightforward: uncertainties in specifying the effects of an operation, the broad diffusion of benefits (and losses) beyond the specific industry or region intended, the difficulty in placing values on property damage and, especially, on loss of life, and the difficulty of assessing long-term effects, if any.

Even in cases where the economic value appears to be assured, the issue may not be simple. The costs of increasing precipitation in order to increase agricultural production, for example, should be compared with the costs of alternate methods of obtaining equivalent increases. Such alternate methods would include crop diversification, selection of strains more resistant to drought, and shifting production to other locations, as well as developing other possible sources of water by means of desalting, re-use, or new dams. It is quite possible that the benefit-to-cost ratio for this example would be very different for the individual farmer, the region, the nation, and the world. Economic evaluations, though necessary, will not dispose of the difficult public-policy problems associated with weather modification.

Legal Aspects The novel problems posed by weather-modification operations have attracted a great deal of attention from legal scholars, and cases have come to trial in the States of New York, Texas, Nebraska, Oklahoma, Washington, California, and Pennsylvania. Analysis of legal issues especially relating to water-resource development was carried out under the leadership of H. J. Taubenfeld[65] of Southern Methodist University and R. J. Davis[67] of the University of Arizona in the late 1960's. Much attention has been given to the problem of who has rights to clouds, but the issue has not been resolved. Clouds that pass over a man's property are potential assets that he may want to put to profitable use, but he cannot *own* the clouds. Analogies with laws regarding water supplies provided by streams, lakes, and wells have been advanced, but these are really not very useful because of obvious differences between the two situations. New law will have to be developed which responds to the issues of (a) the right to use clouds over one's property, (b) the right to enjoy natural precipitation not modified by someone else's intervention, (c) the interest of the public in developing methods of effective weather modification, and (d) the degree of liability incurred when damages are associated with deliberate weather modification.

Traditional procedures of law may not be adequate to deal with the kinds of problems that are characteristic of weather modification. Traditional requirements place on the plaintiff a burden of proof of damages, which it is inevitably difficult for him to provide. And traditional adherence to precedent provides little guidance in a field that involves new techniques and has such broad social ramifications as weather modification. Here, as in so many other aspects of the subject, a degree of innovation appears to be needed. Development of law should keep pace with scientific understanding and technical capability; it would not be desirable now, recognizing present deficiencies in knowledge, to set in place a rigid legal structure controlling weather modification. The greatest need at this time is for legal provisions that will ensure that scientists are able to carry out field research programs necessary to understand the processes associated with precipitation and the effects of deliberate modification. Because of the unity of the atmosphere, laws respecting weather modification should be uniform over the nation. Therefore, primary responsibility for resolving legal issues lies with the federal government.

Environmental Considerations Although most efforts to modify weather have concerned the seeding of clouds, the subject is very much broader. Among the ideas that have been discussed are suppression of evaporation by coating water surfaces, melting ice by spreading carbon dust, raising temperature by growing dark-leaf ground cover, and "glacial" engineering designed to increase runoff from snow fields. To these deliberate efforts can be added such examples of inadvertent modification as the increase of CO_2 content of the atmosphere from burning of fossil fuels, addition of cloud nuclei from industrial and urban sources, and effects of added heat and particulate matter to the atmosphere. These examples demonstrate that weather modification is part of the large group of problems concerned with man's influence on his environment. There must be systematic programs for the evaluation of the impact of weather modification, both deliberate and inadvertent, on the ecology of the major types of natural wilderness areas.

Public sensitivity to these problems has grown rapidly in recent years. We must anticipate that as energy consumption increases and technology advances in coming decades more and more use will be made of the atmosphere and more efforts will be advanced to modify atmospheric processes. Environmental problems will not be *solved* and disposed of; they will have to be *managed* by men and institutions that combine a wide range of technical skills with an understanding of society's needs and desires. Atmospheric scientists will have an important responsibility in this area.

Roles of Government, Industry, and the Universities Government's responsibility in weather modification rests on the fact that the atmosphere is a vital and shared resource, and on the fact that modification inevitably affects the population as a whole in a variety of ways. Only government can be responsible for dealing with the wide range of problems that are intrinsic to weather modification. Obviously, government must develop policy with respect to weather modification, and it must regulate and control. These responsibilities require support of laboratory and field research and of professional investigations into the effects of weather modification on society and on the productivity and ecology of the land.

Industry, by virtue of its flexibility and innovative capability, is especially suited to the development of operational techniques and to the exploitation of the potentialities of weather modification. There is every reason, therefore, to encourage industry to play the primary role in operational programs, subject to regulation by government.

The universities are needed to make crucial contributions to weather modification both in the area of scientific research and in the area of public policy. Universities must train the technicians and the scientists needed for an effective national program in weather modification, and they must maintain the research leadership that is essential to graduate training. Universities also can bring together the specialized capabilities and the objectivity needed in examination of the public-policy issues.

Decision Criteria for Operational Programs If and when hurricanes and other severe storms can be modified, someone will suddenly be saddled with very heavy responsibilities for making operational decisions. Procedures will be needed for ensuring that operational programs for a variety of objectives are in the fullest sense in the public interest. Before those programs are undertaken, there must be assurance that operators are technically competent, all effects of the operation must be known within reasonable limits, all affected parties must be represented in the decision-making process, and adequate provision must be made for liability in case of damages. Because of limited scientific understanding and because of inconsistencies in state laws with regard to weather modification, it is not possible at this time to achieve all these requirements, but a start should be made by the federal government. A uniform system of licensing and mandatory reporting of operational procedures and of results are reasonable initial steps that would be valuable.

International Ramifications Air cannot be confined within national boundaries, therefore many of the problems of weather modification are intrinsically international. According to verbal reports from Sweden, distinctly acidic rainfall in recent years may be responsible for adverse effects on the forests of Scandinavia.[68] The source of the acid has not been identified, although it must be presumed to include the industrial regions of Western Europe, possibly England, and Sweden itself. As industrialization proceeds in the developed and undeveloped nations, this occurrence may be expected to be multiplied manyfold. Many of the problems of which we are beginning to be aware within our own country will assuredly have their counterparts at the international level.

The international dimensions of the problems of weather modification add emphasis to the need for research to advance understanding of the scientific and technical aspects. In addition, discussions should be directed at obtaining international agreement on activities that could seriously affect the weather of regions beyond national borders.

Military applications of weather modification are conceivable and therefore must be considered. In view of the unity of the global atmosphere and the complexities and technical uncertainties of the subject, efforts to use weather modification for military advantage would be likely to be ineffective from the strictly military point of view and would at the same time offer the world increased tensions and new dangers. We know so little about possible responses of the atmosphere that to carry out large-scale interventions into cloud processes or energy-transfer processes would be highly irresponsible. Such efforts also might increase pollution of the atmosphere and ocean under conditions that would make control virtually impossible. In recognition of this danger, at least as it applies to nuclear bomb debris, the United States, the Soviet Union, and other nations adopted the Nuclear Test Ban Treaty of 1963. For these reasons, national policy should seek to minimize the possibility that any country might undertake to modify weather for military purposes.

Responsibility for Public-Policy Decisions The public-policy issues that have been outlined here have much in common with similar issues arising out of such problems as management of the use of the electromagnetic spectrum or nuclear power generation or the use and control of pesticides. Problems like these, in which technological and sociological aspects are intimately combined, are becoming more numerous and more urgent. In these areas wise policy clearly is dependent on scientific understanding and accurate technical information. However, policy decisions will have to be arrived at on the basis of partial information; we cannot wait for scientific certainty. Policy regarding these problems must be developed on the basis of the best information available, with proper provision being made for accommodating to new information.

A recent study by the National Academy of Sciences Panel on Technological Assessment[69] has emphasized the need for development of new institutions within government capable of assessing technical developments and potentialities on a basis that is scientifically sophisticated but broad and humane in perspective. A focus of responsibility was identified as properly residing in the Executive Office of the President with possibly another focus in a joint committee of Congress. The report also proposes a NSF Technology Assessment Division, which would support assessments by universities, independent scholars, nonprofit organizations, and public agencies. These proposed institutions would be very valuable in developing and administering the national program in weather modification.

RECOMMENDATIONS

In view of the potential importance of both inadvertent and deliberate weather modification and the complexity of the subject, a major objective for the 1970's should be

To establish mechanisms for the rational examination of deliberate and inadvertent means for modifying weather and climate.

This objective is regarded as timely and important but as somewhat less urgent than the two objectives stated earlier in this report. A few scientists would place this objective immediately after the first objective (a) because weather modification is more mature as a field of scientific investigation than air quality, and (b) because of the potential importance of ultimately achieving an extensive capability to modify weather. These views were not widely held by study participants.

The recommendations that follow are addressed to the establishment of an effective national program of weather modification which would be responsive to the range of human needs implicit in the subject. Recommendations III-1, III-2, and III-6 concern matters of policy and are addressed to levels of government responsible for making policy decisions. Recommendations III-3, III-4, and III-5 are concerned with the science and technology of weather modification. They are expressed in broad general terms, and more detailed recommendations will be contained in the report being prepared by the Committee's Panel on Weather and Climate Modification, which was referred to earlier in this chapter.

Recommendation III-1 In order that public-policy issues associated with weather modification may be handled effectively and wisely, *responsibility for examination of public-policy issues and for developing policy initiatives should be lodged in the Executive Office of the President.*

These functions probably should be carried out under an appropriately strengthened Office of Science and Technology, closely coordinated with and supported by the observational and research capabilities of the federal agencies and the scientific community. Social scientists, lawyers, political scientists, and public officials must play central roles in policy development.

Assignment of first priority to this recommendation reflects the overriding importance of the public-policy issues in this field and the fact that existing procedures for the determination of policy will not be adequate to the needs of the 1970's.

Recommendation III-2 In order to make possible a national program in weather modification capable of responding to a variety of needs, *principal administrative responsibility for the national program in weather modification should be assigned to the National Oceanic and Atmospheric Administration.*

This recommendation will require that the NOAA budget include substantial funds allocated to weather modification. In addition to carrying forward a vigorous research program, NOAA should be responsible for coordinating the mission-oriented research supported by other agencies with the dual objectives of enhancing the total effort on the central scientific problems and assisting individual agencies to make their programs as effective as possible.

Recommendation III-3 In order to learn what effects on global weather and climate are being produced by the various forms of environmental pollution, *a global network of monitoring and research stations should be established under international auspices.*

The sites selected for stations must be located away from sources of pollution, and there must be reasonable assurance that they will remain so. Monitoring programs should be started promptly at approximately four stations for CO_2 and aerosol concentrations, electrical conductivity, and solar radiation. These elements are directly relevant to the understanding of changes in climate, and they are amenable to unambiguous, reliable, and standardized measurement. A regular program of calibration must be instituted. Research stations should be established at 10 or more sites for the purposes of carrying out research on global pollution and its effects and for testing new techniques for measuring concentrations of pollutants.

The monitoring program recommended here strongly overlaps the global monitoring programs recommended in Chapters 2 and 3. Although the three observation programs are much the same, the objectives are different, and the total urgency of a global monitoring system must be considered additive.

In considering the role of the atmospheric sciences in the regional and global aspects of air quality and climate change, the Committee has focused on potential interactions between air pollutants and weather and climate processes. Recommendations I-3, II-2, and III-3, which concern global monitoring of pollutants and climate change, reflect this focus. The broader ecological aspects of global pollution by materials such as heavy metals, trace gases, and radionuclides, although of great importance, are not addressed explicitly in this report. In these problem areas atmospheric research must continue in close cooperation with medical, biological, ecological, economic, and political research. Such cooperation has been greatly advanced by the International Biological Program and, more recently, by the planning work of ICSU/SCOPE and the United Nations agencies in preparation for the United Nations Conference on Man and His Environment in 1972.

Recommendation III-4 In order to carry out large field experiments addressed to critical scientific problems, *a new or an existing major laboratory should be assigned responsibility and resources for developing the necessary scientific and technical capability and for implementing field research programs including coordination with universities and the federal government for planning and participation in field efforts.*

The critical need is for facilities and manpower of adequate capability to carry out definitive field research programs in weather modification. The laboratory could be established under any of the following administrative arrangements: a new laboratory within the

Environmental Research Laboratories of NOAA, expansion of existing groups within NCAR, or development of weather-modification capability within a university-affiliated laboratory or institute supported by one of the agencies active in weather modification. The program should be staffed by a thoroughly competent group of approximately 100 atmospheric scientists and technicians, and stable support should be provided for a minimum period of 10 years.

The following are examples of the crucial problems that require the capabilities of a major laboratory: (a) further systematic investigation of the dynamic effects that can be produced by cloud seeding in order to determine whether hurricane winds can be reduced in severity and their paths changed and, if so, the design of operational programs to begin to carry out this objective; (b) investigation of the application of cloud-seeding techniques to modification of squall lines, thunderstorms, and other severe storms; (c) further investigation of the atmospheric characteristics that distinguish positive effects of cloud seeding from negative; and (d) determination of the optical properties of atmospheric particles in order to determine whether increased global concentrations of particles will produce cooling or warming. The laboratory also should carry out studies relating to critical aspects of the public-policy issues identified earlier. Social scientists, statisticians, lawyers, and political scientists should work closely with atmospheric scientists on these problems.

The first three of the programs listed above represent the heart of efforts to modify weather by deliberate means. They have been much worked on, but efforts to date have been too small and weakly supported for the results to be definitive.

Recommendation III-5 In order to advance the development of effective weather-modification techniques more rapidly, *substantially greater research efforts in cloud physics and weather modification should be carried out in university, private, and government laboratories.*

A large number of research programs are amenable to treatment in such laboratories. It is particularly important that attention be devoted to the development of numerical models of various atmospheric phenomena ranging from cumulus clouds to tropical cyclones. This goal is needed not only for the purpose of developing procedures for the deliberate modification of the weather, but it is closely related to the development of the short-range prediction capability recommended in Chapter 2.

The largest available computer should be made available to at least one numerical modeling group, and increased support should be provided to universities for training and research in this area. Numerical model work should be coupled to field investigation of clouds.

The following represents other examples of the kinds of undertaking that can be successfully dealt with by a small research group: (a) development of instrumentation for measuring cloud nuclei and cloud particles; (b) laboratory and field experiments to determine specific effects of artificial nucleation on precipitation and other weather phenomena; (c) laboratory and field experiments to extend understanding of the charge-separation mechanism; and (d) determination of the value of remote-sensing techniques for observing changes in atmospheric conditions resulting from weather-modification experiments.

The following additional action would provide a significant strengthening of the national program and would contribute to international understanding and cooperation.

Recommendation III-6 In order to safeguard the life-sustaining properties of the atmosphere for the common benefit of mankind, *the U.S. Government is urged to present for adoption by the United Nations General Assembly a resolution dedicating all weather-modification efforts to peaceful purposes and establishing, preferably within the framework of international nongovernmental scientific organizations, an advisory mechanism for consideration of weather-modification problems of potential international concern before they reach critical levels.*

CONCLUSION

These recommendations cover the most urgent of the programs that should be carried out in coming years. Our attention has been deliberately restricted, rather than catholic, in order to place a sharp focus on the elements that are expected to be most crucial in meeting important human needs. We believe that weather modification should not be viewed as an optional program that conceivably might be left to private initiative or to the ebb and flow of federal budgets but rather as an inevitable activity that the government must meet responsibly and intelligently.

5
WEATHER DANGERS AND DISASTERS

An elemental force is ruthlessly frank.
JOSEPH CONRAD

Each year weather disasters cause the loss of many lives and the destruction of crops, structures, and other property. A major responsibility of the atmospheric sciences has always been to anticipate these events and to minimize the resulting losses and dislocations. In the next decade the dangerous and damaging effects of weather can be further substantially reduced by application of what we know about the atmosphere. This is the common theme of the recommendations of the three preceding chapters of this report. In this chapter some of the specific actions are identified that can result in anticipating dangerous weather situations, reducing property loss, and saving lives.

In some weather situations, meteorologists are of greatest service by detecting severe weather systems and by warning individuals and communities in their path. In other cases, it may be possible to forecast weather disasters hours or even days in advance and to provide warnings that permit evacuation of people and protection of property. Finally, with certain specific types of severe storms, there may be possibilities of modifying the weather to prevent loss. Each different type of weather disaster has its own characteristics that determine how its destructive effects can best be minimized—by detection, tracking, and warning; by forecasting and warning; or by modification of the storm itself.

Severe weather covers a very wide range of time and space scales. Local, intense storms that affect only a small area do not usually have a long lifetime. In general, such systems cannot be forecast far in advance. The meteorologist must therefore concentrate his efforts toward detecting the storm, forecasting its motion for periods of a few minutes up to a few hours, and issuing timely warnings. Examples of such local severe weather are tornadoes, severe thunderstorms, and squall lines. Even in the case of such small-scale systems, however, the possibility exists of modifying the weather: for example, in the Soviet Union potential hail-producing thunderstorms are seeded by antiaircraft shells or rockets carrying lead iodide in an effort to reduce their severity and limit crop loss.

Other types of severe weather features are somewhat larger in areal extent and have lifetimes that are measured in tens of hours or in days. In such cases the meteorologist's task is to prepare the best possible forecast and then to update his forecast continually in order to provide the best possible warning to the affected populace. Modification of these larger storms by artificial techniques may or may not be feasible, depending on the type of storm under consideration.

Weather disasters may also result from unusually mild weather associated with very large-scale weather systems that may persist for days or even weeks. Examples include prolonged drought or the existence of stagnant air masses, which may lead to severe pollution in urban areas. At present, there is nothing that meteorologists can do to modify these huge weather systems, which may cover hundreds of thousands of square miles. However, as has been pointed out in Chapter 2, it may be possible within the next decade to provide accurate notice of a significant break in the weather as much as a week or ten days in advance.

For each of the following weather phenomena substantial reduction in danger and losses should be possible within the next decade. The dangerous characteristics and the necessary research and development are indicated for each phenomenon. Phenomena are discussed in order of increasing time and space scale, beginning with storms that are short-lived and small in areal extent and progressing to long-lasting phenomena that cover large regions.

TORNADOES

The tornado is the most vicious and destructive of all small-scale storms. Hundreds of people lose their lives because of tornadoes each year, and many more are made homeless. Often these storms strike with little or no warning, and the public has little time to seek protection or get out of the storm's way. Tornado funnels typically have a lifetime of a few minutes to half an hour. They may vary in width from 10 to 1000 yards. Tornadoes often occur in groups: when one funnel is observed, others may develop in the vicinity.

At present, the National Weather Service issues forecasts delineating regions, perhaps a hundred miles across, within which tornadoes may occur. When specific storms are identified, local emergency alerts and warnings are issued. Although these types of forecast are moderately successful, there is little likelihood that the formation and location of individual funnel clouds can be predicted. Additional research on the structure and dynamics of tornadoes is necessary. Given the capricious nature of these storms and our present inadequate understanding of their physics, the development of techniques for modifying tornadoes must be regarded as a distant goal.

There is every hope, however, that surveillance systems can be devised that will detect tornadoes once they have formed. If the full capabilities of geostationary satellites, radar, sferics detectors, and other remote-sensing techniques are brought to bear as a component system, we should be able to detect tornadoes, track their motion, and prepare short-term warnings capable of saving many lives. This will require rapid and effective communication to the public. An automatic signal, transmitted through the electrical power supply, might be used to switch on radios or television sets in threatened areas. Warnings to the public could then be issued using dedicated or commercial television or radio channels, which can also be used to signal "all clear" when the storm's danger has passed.

The development of an accurate tornado detection, tracking, and warning system will require a coordinated research and development effort. The Committee strongly recommends that such a program be set up as soon as possible and that a pilot warning system should be placed in operation for the 1973 tornado season at some appropriate location in the Midweat. This could be incorporated in the field programs recommended in Chapters 2 and 3, if suitably located.

SEVERE THUNDERSTORMS AND SQUALL LINES

These storms are less severe than tornadoes but occur much more frequently. Cumulatively, they result in hundreds of millions of dollars in property damage each year. Individual isolated thunderstorms are often 5 to 10 miles in diameter and may have a total lifetime of 1 or 2 hours. Frequently, thunderstorms occur in lines that tend to move at a fairly regular rate. In such cases, the problem of the forecaster may be somewhat easier, since a squall line may remain active for many hours, and its time of arrival can be forecast with steadily increasing accuracy as it approaches.

As is the case with tornadoes, the location and movement of individual thunderstorms or squall lines are difficult to forecast *before* the weather system has actually formed. Thus, the attention of the atmospheric scientists is again primarily directed to detecting these storms; preparing short-term forecasts of their motion; and disseminating warnings. The Local Weather Watch recommended in Chapter 2 will be of prime importance in providing timely warnings of such weather systems. It has been recommended that the Local Weather Watch be developed in two pilot programs, one in an urban and the other in a rural area. Success of these pilot programs, and extension of the Local Weather Watch system to additional areas, would provide an optimum method for reducing loss of life and damage to property from severe thunderstorms and squall lines.

A strong research program must also be continued that will seek to provide a better understanding of the physics of these storms and of the way they can be modified. Cloud-seeding techniques have already been used in an effort to alter certain types of severe thunderstorms, such as those that produce hail, and efforts toward thunderstorm modification should continue. In the long run, weather modification may provide the best procedures to cope with local severe thunderstorms and squall lines.

HURRICANES

The hurricane is one of nature's most destructive storms. Hundreds of lives were lost in Hurricane Camille in 1969 and over a billion dollars was lost in property damage. Were it not for the present U.S. hurricane warning service, the loss of life could have reached thousands, as was illustrated by the tragic loss of life in the recent East Pakistan hurricane catastrophe of November 1970. Yet Camille was by no means a unique example. A few years earlier, in 1965, Hurricane Betsy caused $1.5 billion damage, and each year hurricanes of "normal" intensity take a heavy toll.

The typical hurricane has a diameter of 50 to several hundred miles and a lifetime measured in days. It can be detected by weather satellites or by reconnaisance aircraft. Once its existence is known, it is now possible to track it accurately, and the ability to forecast the true motion of a hurricane is gradually but steadily increasing. The effectiveness of the hurricane warning service in the United States is illustrated by Figure 2, which shows that the loss of life due to these storms has been steadily decreasing. This is true despite the fact that property losses caused by hurricanes have risen as cities and industries have developed along exposed coasts. Similar statistics show that the number of deaths caused by tornadoes has decreased in the past quarter century, although the number of reported tornadoes has greatly increased.[44]

Hurricanes are an international problem. They originate and grow to maturity over the oceans of the globe, often far from any weather-observing station. In recent years, a totally new capability to monitor the ocean in order to detect and track these storms has been provided by weather satellites, especially the ATS synchronous satellites. Satellite observations are also used effectively to plan reconnaissance flights of manned aircraft in probing storm centers.

During the next decade, additional funds must be allocated for further improvement of the already excellent hurricane forecast and warning system. Major improvements must be made in the way forecasts are disseminated to the public: in Hurricane Camille, for example, many lives were lost needlessly when individuals refused to heed repeated requests to evacuate vulnerable coastal areas. Again, the possible use of controlled variations of voltage in public power supplies, as a means of activating radio and television sets for emergency messages, should be seriously considered.

In the long run, however, the most effective way to cope with hurricanes may be by their modification using aircraft seeding techniques. Important experiments were conducted during the 1969 hurricane season, when the maximum winds at 12,000 feet in Hurricane Debbie were twice observed to be substantially lower after seeding with silver iodide.[70] Calculations made by a simple numerical hurricane model indicated wind decreases at the sea surface as a consequence of seeding. These results, while promising, are not yet conclusive. Substantial additional resources should be devoted to further develop this series of experiments. In order to obtain the necessary data promptly, the possibility of conducting the tests in the Pacific as well as the Atlantic should be considered. In addition, funds

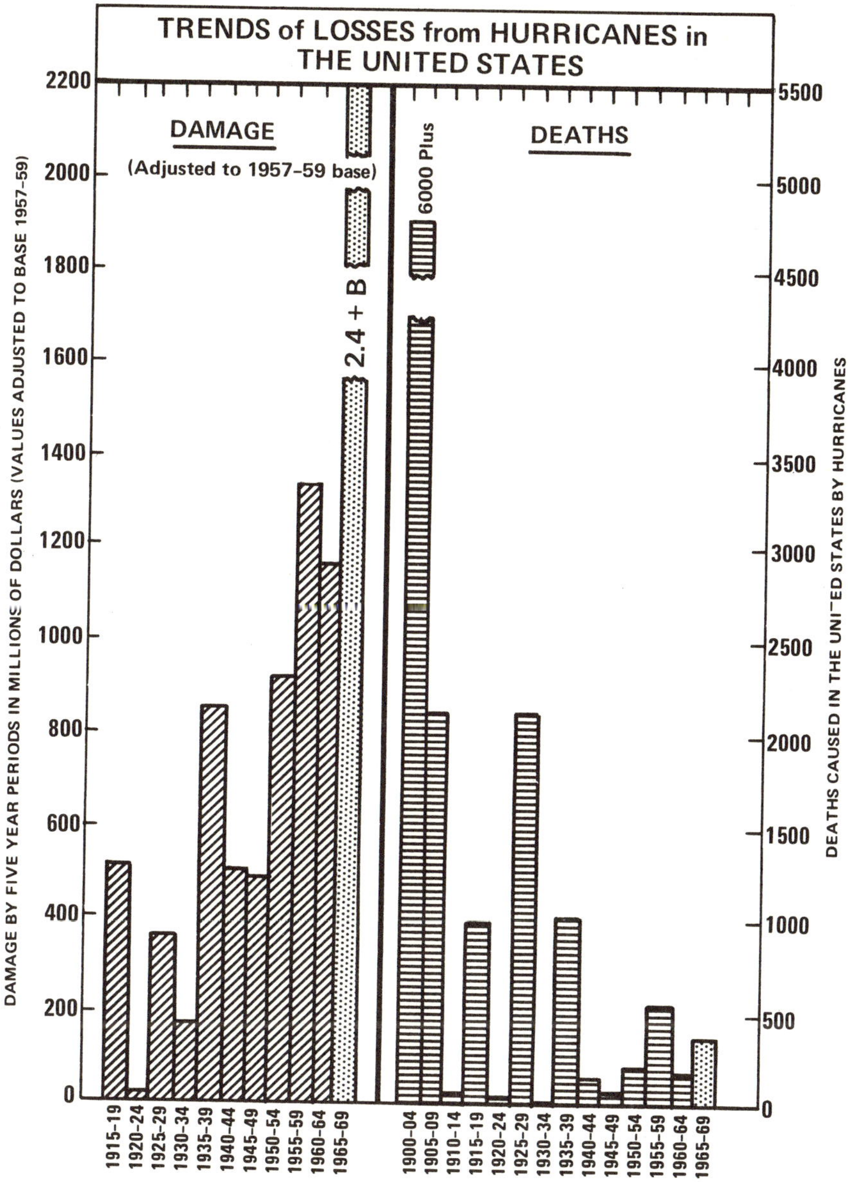

FIGURE 2 Damage and death statistics indicating marked reductions in loss of human life through improved warning services. The growing vulnerability of certain portions of the country is also revealed by the increased property losses. (After Gentry, Reference 70.)

should be allocated to improve our capability to simulate hurricanes on high-speed computers. Numerical simulation and field experimentation must be carried out as coordinated efforts.

POLLUTION DISASTERS

Until atmospheric pollution has been brought under control by the reduction of harmful effluents, the possibility of local disasters occurring as a result of severe pollution cannot be discounted. Such occurrences can be of two different types. An accidental release of some pollutants such as toxic or radioactive chemicals in storage or transit, or from an industrial plant, will result in dispersal over downwind regions and may require the evacuation of people and restriction on the use of agricultural products from the affected areas. If such disasters were to occur, the operational meteorologist must provide information concerning the areas in which hazards might become severe. The Local Weather Watch, recommended in Chapter 2, would provide the observational data and information-dissemination system to make such predictions more accurate and useful. Remote sensing of pollution may be especially useful for this purpose.

Pollution disasters also have arisen as a result of persistent weather systems in which stagnant air masses allow the local accumulation of pollutants in an urban or other industrial area. This typically occurs when a large anticyclone is present over a particular region. If the anticyclone is associated with a low-level inversion (cold air near the surface and warm air aloft) pollutants tend to be trapped near the ground. In these cases the concentration of noxious or toxic airborne materials may increase over a period of days and may reach unacceptable levels.

The motion of anticyclones can be predicted for one or two days using high-speed computers and other standard forecasting techniques. If the forecast is for continued stagnation, local authorities can plan a schedule of emergency actions to reduce the release of additional pollutants to the atmosphere. The accuracy and extent of such forecasts should be materially increased during the next decade by the World Weather Program. The recently established Stagnation Warning should be upgraded as promptly as technology permits by coupling advanced observational capability with numerical prediction capability.

CONCLUSION

The above list of weather disasters is by no means complete. There are also severe hail storms, which can cause major damage to buildings, vehicles, and crops; local snow storms, which can cripple an urban area, especially when they strike without warning; larger-scale blizzards, often whipped by gale winds, which can isolate entire communities and interrupt vital transportation; severe drought or floods, which can cause serious economic loss over substantial areas; and forest fires, which can destroy timber over substantial areas. In each of these situations, the meteorologist has a key role to play.

For the cases discussed in this chapter, excellent opportunities now exist to reduce or to anticipate dangers and losses due to weather events. Therefore, a major objective in the atmospheric sciences during the 1970's should be

To reduce substantially human casualties, economic losses, and social dislocations caused by weather.

This objective follows the objectives listed earlier because its implementation depends upon the programs recommended in Chapters 2, 3, and 4. It seems likely that in the next decade the atmospheric sciences will contribute very significantly to man's efforts to contend with the hostile aspects of his environment.

6 THE RESOURCES REQUIRED

Wisdom lies in masterful administration
of the unforeseen.

ROBERT BRIDGES

After a decade of technical development and growing scientific strength in the atmospheric sciences, we are entering a period in which these sciences can be more effectively brought to bear on human needs than has been true heretofore. The opportunities for improving and extending weather predictions to longer periods, for developing procedures for maintaining equitable levels of air quality, and for protecting people and property from the ravages of capricious weather events have all been pointed out in this report. To realize these opportunities, a continuing commitment of manpower, facilities, and money to support both operational activities and research and development is required.

RESOURCE ALLOCATIONS DURING THE 1960'S

The dramatic commitment of the nation to strengthened research in the atmospheric sciences during the early 1960's is portrayed in Figure 3. During the four years following 1960, the annual research expenditures in the atmospheric sciences rose from $37 million to $222 million (Interdepartmental Committee for Atmospheric Sciences

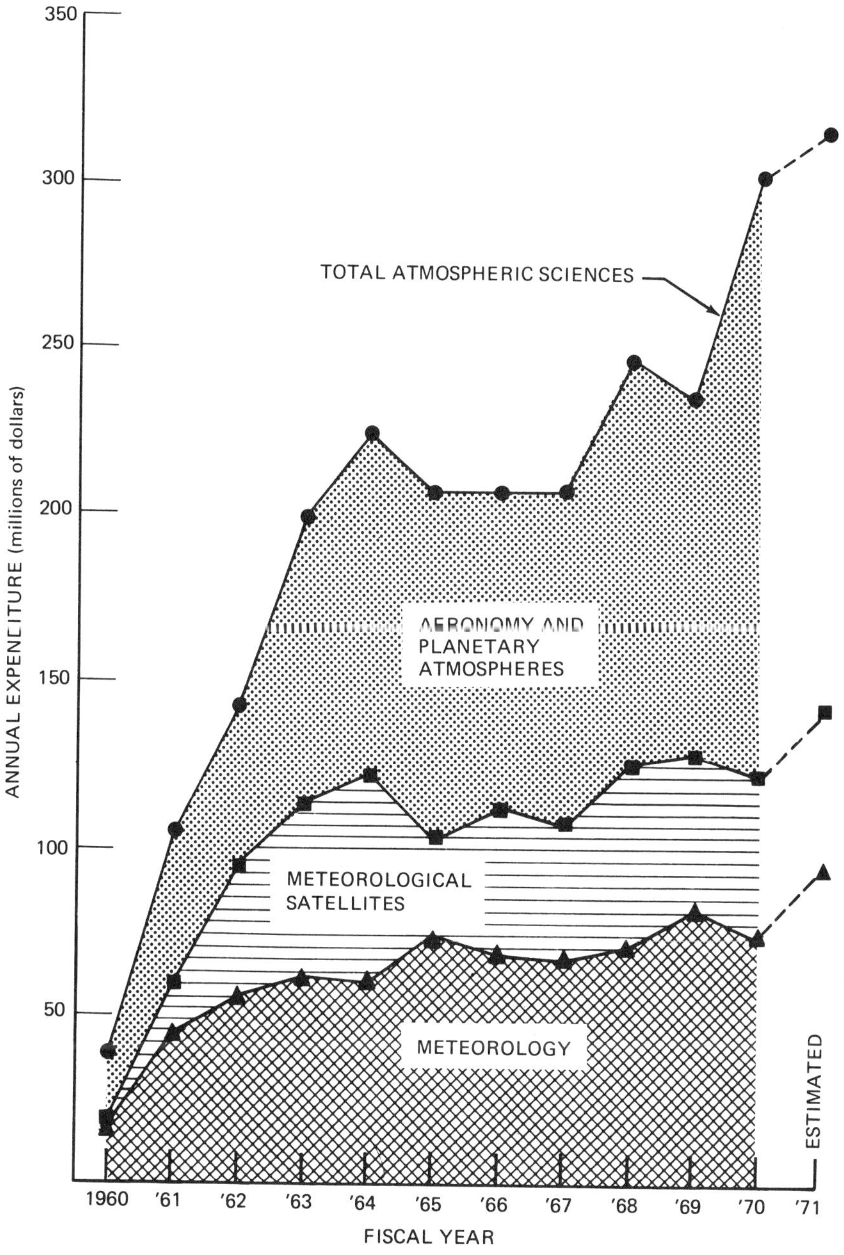

FIGURE 3 Federal expenditures for atmospheric-sciences research during the 1960's. Total expenditures during 1960–1969 were aeronomy and planetary atmospheres, $875 million; meteorology, $509 million; meteorological satellites, $406 million.

figures). About half of the total investment has been directed to the general fields addressed in this report; and this is represented with reasonable accuracy by the sum of "Meteorology" and "Meteorological Satellites" in Figure 3. The other half of the total investment for the decade was in "Aeronomy and Planetary Atmospheres."

The dominant message conveyed by Figure 3 is that support for research increased strongly during the first four years of the decade, and that support for meteorological research has increased very little since fiscal year 1965. In the period since 1965, support for meteorology and meteorological satellites has increased on the average at an annual rate of 4 percent per year, substantially less than the rate of increase necessary to offset inflation and the costs due to increasing sophistication in facilities and instrumentation. As a consequence, it has become more and more difficult to mount large research programs and to acquire needed large facilities. The increase in aeronomy and planetary atmospheres in 1970–1971 reflects the preparation for major programs of NASA in planetary exploration during 1971, 1972, and 1973, largely for procurement of space platforms.

Many university programs are in special jeopardy. During the first 7 years of the decade allocations to universities increased at about 10 percent per year and reached their highest level in fiscal year 1967. In Table 1, the situation is indicated as it has developed since 1967. Support for university research has dropped by $7.6 million in that period, largely because of severe reductions in NASA and DOD university programs, despite the valiant efforts on the part of NSF to increase its funding for worthy basic research programs. Separate figures for meteorology are not available, but they probably represent about half of the total for each of the years. The decrease in university research support may be compared to an increase in the same period of 13 percent in the total investment in meteorological research and development.

TABLE 1 Support ($ Millions) for Atmospheric Sciences and for University Research in the Atmospheric Sciences, 1967 and 1970[a]

	Atmospheric Sciences	Universities
FY 1967	208	41.7
FY 1970	303	34.1
Change, %	+45	–18

[a]Budgetary data obtained from ICAS Reports Nos. 10 and 14 and supplemental ICAS information.

The 18 percent cut in basic atmospheric research in the universities, during the immediate past period of increasing costs, has resulted in a critical shortage of research support for young scientists. Established university research programs, which provide the primary support for graduate students, are being cut back. Offsetting this cut to an uncertain but, we think, small extent are the opportunities provided by the National Center for Atmospheric Research (NCAR) and by government laboratories for university scientists to utilize research facilities and to participate in large research programs.

The level of support for atmospheric research in the past several years, especially for university research, must be substantially raised in order that the pressing needs of the 1970's can be met.

We anticipate no severe shortage in manpower; graduate degrees have increased at about 7 percent per year in the past decade, and recruitment of engineers has become an effective channel for provision of technical manpower. Specific areas are of some concern. There will be need for substantial increase in the number of specialists in numerical modeling and in atmospheric chemistry. There will also be need for increased numbers of engineers working on application of remote sensing to atmospheric problems, but we see no need for specialized educational programs in this area.

The continuing educational programs initiated during the 1960's by the American Meteorological Society have also contributed to the steady growth of professional interest in the field. This program is of special importance for long-term recruitment.

Beyond qualified manpower and fiscal support, the primary requirements for successful research and development are major facilities, particularly computers, aircraft, and radars. The expenditure of approximately $50 million for major facilities during the 1960's may be listed as follows:

Computer facilities	$24.0 million
NCAR (other than aircraft and computers)	16.5 million
Radar and other remote probing*	3.5 million
Aircraft facilities†	6.0 million
TOTAL	$50.0 million

*This figure represents direct expenditures of atmospheric-sciences funds. In addition, much of the research progress in this field has depended on extensive facilities, funded separately by DOD and NASA, at Wallops Island.

†This figure includes the purchase and instrumentation of NCAR's flight facility, but only instrumentation of NOAA's Research Flight Facility. Acquisition was by transfer.

TABLE 2 NAS/CAS Recommendations for New Research Funding in the Atmospheric Sciences during 1970–1979

			New Money ($Millions)	
Atmospheric Sciences Program	NAS/CAS Recommendations	NAS/CAS Priority Ranking	For Research Operations	For Research Facilities
Weather prediction	Full implementation of GARP	I-1	75	Computer facilities
	Local Weather Watch	I-2	4	8[a]
	Dynamics and modeling of climate	I-3, II-3	25	Computer facilities[b]
	Mesoscale research	I-4	15	10[c]
Air quality	Urban models and experiments	II-1	30	Computer facilities
	Remote sensing and research	II-1, III-5	25	13[d]
	Global air quality	II-2	20	Computer facilities[b]
	Regional models and measurements	II-3	16	Computer facilities
Weather and climate modification	Assignment of direction and administration of program	III-1, III-2, and III-6	– – –	– – –
	Weather-modification laboratory	III-4, I-4	35	8[e]
	Cloud dynamics and physics models and measurements	III-5, I-4	25	Computer facilities 4[f]
Consolidated computer requirements	In support of I-1, I-3, II-1, II-2, II-3, III-3, and III-5	–	–	140
TOTALS			270	183

[a]Facilities include satellite and ground-based sensor communication links, data processing, and data transmission system. Research operations budget is for one demonstration only and does not include follow-on funding for further development or demonstration.
[b]These recommendations assume international cooperation and funding of global programs with the U.S. funding $10 million of a total cost (for the decade) of $30 million for climate monitoring and $20 million of a total cost of $60 million for meteorological aspects of global air-quality monitoring.
[c]Facilities include new aircraft for mesoscale research, probably associated with an existing flight facility and instrumentation costs only for replacement of NOAA's Research Flight Facility aircraft.
[d]Facilities for an expansion of remote-sensing research activities in NOAA and in several university groups. The programs recommended under II-1 and III-5 depend strongly on continuing access to other major radar facilities, such as Wallops Island and Millstone Hill.
[e]Facilities expenditures represent estimated costs of refurbishing or extending physical facilities of an existing laboratory suitable for research and field facilities necessary for complex weather-modification experiments. Operational costs and facilities are estimated on the

In addition to these major facilities, numerous laboratory and office facilities, temporary field measurement facilities, and specialized networks have been constructed. Their costs are accounted for as research expenditures and are not identifiable as facilities. (For a partial list of meteorological facilities see Reference 71.)

The outstanding needs for major facilities to support the recommended programs of this report are the following:

1. Computer facilities must be upgraded to fourth- and fifth-generation capacity and speed at at least three major centers.
2. The need to develop techniques for the remote probing of the lower atmosphere and clouds requires the development of several active university groups and further strengthening of NOAA's Wave Propagation Laboratory.
3. Weather modification and cloud physics and dynamics studies require the development and acquisition of modern radar equipment.
4. The aircraft of NOAA's Research Flight Facility must be replaced. In addition, a flight facility should be assembled for mesoscale research on clouds and atmospheric structure.
5. A National Weather Modification Laboratory or Center should be provided.

In addition to these major, identifiable facilities, the programs recommended in this report will require offices, laboratories, shops, computer support, and instrumentation associated with university, industrial, and governmental research laboratories. These costs are included as regular research expenditures.

TO THE 1970'S

The estimated costs for research and facilities required by the recommendations of this study are summarized in Table 2. Estimates are

---

Footnote to Table 2—Continued.

basis of a research and facilities staff of approximately 100 people operating primarily in a field-measurements mode.

[f]Facilities costs are estimated for extension or refurbishing of NOAA's present Wave Propagation Laboratory physical facilities to accommodate a 50 percent increase in the remote-sensing research and development at that laboratory. Operational costs of approximately $1.3 million per year are allocated for this activity. The balance of $12 million operational costs for the decade are allocated for increased research in university and private laboratories on problems of cloud dynamics and physics as well as individual projects on measurements systems.

in 1970 dollars. Emphasis has been placed on the aggregate expenditures for the decade, even though the recommended early scheduling for the Local Weather Watch, the urban air-quality measurements, and facilities acquisitions should result in a period of relatively rapid growth in the early years of the decade, followed by a period of slower growth.

In estimating program costs it has been necessary to examine the details of manpower and facilities needed to provide the minimum capability to achieve the objectives defined in this report. In several instances existing planning documents, e.g., GARP reports and agency analyses of computer requirements, have provided guidance for these analyses. In other areas, e.g., Local Weather Watch, it has been necessary to consider each element of the entire system and to estimate associated costs independently.

The estimated costs have been tabulated separately for each recommendation. The aggregate expenditures for the decade amount to $119 million for weather prediction research, $91 million for air-quality research, and $60 million for weather and climate modification research—a total of $270 million. As the third column of Table 2 indicates, some of the recommended programs are mutually supporting and overlapping, so that the budgetary allocations among these programs are somewhat arbitrary. In support of these activities, about $183 million will be required for major facilities during the decade. Of this amount, $140 million is to be allocated to the acquisition of increased computer capacity and speed.

In making estimates of costs associated with international programs, i.e., Global Air Quality and Climate Monitoring, we have assumed international funding and a U.S. contribution of one third of the total cost.

The actions recommended in this report are of course dependent on ongoing operational and research programs. In some cases, they are direct extensions of existing or planned programs. Cost estimates apply to the additional facilities and research required. The possibilities of reprogramming current budgets and of the shared use of operational computers should be examined by appropriate government agencies. However, it seems clear that only a small part of the needed resources can be met by these routes, and that substantial increases in agency research budgets will be necessary in order to realize the proposed objectives. The Committee considers the estimated total cost of $453 million to be low compared to the benefits that can be achieved in the next 10 years.

7

SUMMARY OF OBJECTIVES AND RECOMMENDATIONS

The committee recommends, and ranks according to priority, the following objectives and actions for the Nation:

OBJECTIVE I *To extend the capability for useful prediction of the weather and atmospheric processes.*

This objective is given the highest priority because important improvements in prediction are attainable and feasible and because prediction provides the base that supports the remaining objectives of this report. The ability to modify or to control atmospheric processes, or to anticipate dangerous weather conditions, depends directly on improving predictive capability. Recent achievements in science and technology make this objective timely and feasible.

In order to achieve this objective the following actions are proposed:

Recommendation I-1 In order to extend the range of useful prediction into the one- to two-week period, *the international Global Atmospheric Research Program (GARP) should be vigorously supported, and adequate resources should be provided promptly for necessary GARP experiments.*

Recommendation I-2 In order to determine the potential value of short-period extrapolations of local weather, *a pilot Local Weather Watch should be conducted for a selected urban area and a selected rural area that would provide real-time visual presentations of critical weather features.*

Recommendation I-3 In order to assess the possibilities of anticipating changes in climate and of making useful seasonal predictions, *increased efforts should be directed toward understanding the dynamics of climate.*

Recommendation I-4 In order to advance understanding of mesoscale phenomena and to improve ability to forecast these phenomena, *emphasis should be placed on research on fronts, jet streams, organized convection, and other phenomena of mesoscale.*

OBJECTIVE II *To contribute to the development of the capability to manage and control the concentrations of air pollutants.*

This objective is assigned high priority in view of the urgency for effective and equitable management of waste disposal to the atmosphere so as to avoid harmful effects on human health, ecological systems, and potential climatological changes.

In order to achieve this objective the following actions are proposed:

Recommendation II-1 In order to develop reliable methods to assist in the assessment of alternative urban pollution control programs, (a) *development and testing of simulation models of the urban source–atmosphere system should be accelerated,* and (b) *a major field program of at least two years' duration, designed to measure in detail the fate of pollutants emitted from an urban area, should be carried out.*

Recommendation II-2 In order to assess the potential effects of the global distribution of pollutants, *an effective system for global monitoring of atmospheric contaminants should be established promptly.*

Recommendation II-3 In order to understand the processes that determine air-pollution concentrations on a regional scale, *the chemical composition of precipitation and the dry fallout should be measured for one or more selected areas and periods.*

OBJECTIVE III *To establish mechanisms for the rational examination of deliberate and inadvertent means for modifying weather and climate.*

The consequences of deliberate or inadvertent modification of weather phenomena may be so far-reaching that we must know much more specifically what is possible and what is beyond human efforts. Demonstration of significant weather modification must be expected to trigger profound ethical, social, economic, and political problems, which should be given appropriate attention.

In order to meet this objective the following actions are proposed:

Recommendation III-1 In order that public-policy issues associated with weather modification may be handled effectively and wisely, *responsibility for examination of public-policy issues and for developing policy initiatives should be lodged in the Executive Office of the President.*

Recommendation III-2 In order to make possible a national program in weather modification capable of responding to a variety of needs, *principal administrative responsibility for the national program in weather modification should be assigned to the National Oceanic and Atmospheric Administration.*

Recommendation III-3 In order to learn what effects on global weather and climate are being produced by the various forms of environmental pollution, *a global network of monitoring and research stations should be established under international auspices.*

Recommendation III-4 In order to carry out large field experiments addressed to critical scientific problems, *a new or an existing major laboratory should be assigned responsibility and resources for developing the necessary scientific and technical capability and for implementing field research programs including coordination with universities and the federal government for planning and participation in field efforts*

Recommendation III-5 In order to advance the development of effective weather-modification techniques more rapidly, *substantially greater research efforts in cloud physics and weather modification should be carried out in university, private, and government laboratories.*

Recommendation III-6 In order to safeguard the life-sustaining properties of the atmosphere for the common benefit of mankind, *the U.S. Government is urged to present for adoption by the United Nations General Assembly a resolution dedicating all weather-modification efforts to peaceful purposes and establishing, preferably within the framework of international nongovernmental scientific organizations, an advisory mechanism for consideration of weather-modification problems of potential international concern before they reach critical levels.*

OBJECTIVE IV *To reduce substantially human casualties, economic losses, and social dislocations caused by weather.*

This objective follows the objectives listed earlier because its implementation depends upon the programs recommended under Objectives I–III. Its implementation also requires a commitment by the Nation to reduce the dangers and losses due to weather events.

REFERENCES

1. *Physics of the Earth in Space: The Role of Ground-Based Research,* 1969: Report of a Study by the Committee on Solar-Terrestrial Research of the Geophysics Research Board, National Research Council (National Academy of Sciences, Washington, D.C.).
2. *Useful Applications of Earth-Oriented Satellites: Report of the Central Review Committee,* 1969: Summer Study on Space Applications by the Division of Engineering, National Research Council (National Academy of Sciences, Washington, D.C.).
3. *Physics of the Earth in Space: A Program of Research 1968–1975,* 1968: Report of a Study by the Space Science Board (National Academy of Sciences, Washington, D.C.).
4. *The Atmospheric Sciences 1961–1971. Vol. I, Goals and Plans; Vol. II, Summaries of Planning Conferences; Vol. III, Goals and Plans for Aeronomy,* 1962: A Report to the Special Assistant to the President for Science and Technology by the Committee on Atmospheric Sciences, NAS-NRC Publ. 946 (National Academy of Sciences–National Research Council, Washington, D.C.).
5. *Interaction between the Atmosphere and the Oceans,* 1962: Report of the Joint Panel on Air–Sea Interaction to the Committee on Atmospheric Sciences and the Committee on Oceanography, NAS-NRC Publ. 983 (National Academy of Sciences–National Research Council, Washington, D.C.).
6. *Weather and Climate Modification: Problems and Prospects. Vol. I, Summary and Recommendations; Vol. II, Research and Development,* 1966: Final Report of the Panel on Weather and Climate Modification to the Committee on Atmospheric Sciences, NAS-NRC Publ. 1350 (National Academy of Sciences–National Research Council, Washington, D.C.).
7. *The Feasibility of a Global Observation and Analysis Experiment,* 1966: A Report of the Panel on International Meteorological Cooperation to the Committee on Atmospheric Sciences, NAS-NRC Publ. 1290 (National Academy of Sciences–National Research Council, Washington, D.C.).
8. *Atmospheric Exploration by Remote Probes. Vol. 1, Summary and Recommendations; Vol. 2, Proceedings of the Scientific Meetings of the Panel on*

Remote Atmospheric Probing, 1969: Final Report of the Panel on Remote Atmospheric Probing to the Committee on Atmospheric Sciences (National Academy of Sciences–National Research Council, Washington, D.C.).

9. R. S. Lindzen, 1966: On the theory of the diurnal tide. *Mon. Weather Rev., 94,* 295–301.
10. J. R. Booker and F. P. Bretherton, 1967: The critical layer for internal gravity waves in a shear flow. *J. Fluid Mech., 27,* 513–539.
11. K. Ooyama, 1969: Numerical simulation of the life cycle of tropical cyclones. *J. Atmos. Sci., 26,* 3–40.
12. S. L. Rosenthal, 1970: Experiments with a numerical model of tropical cyclone development. *Mon. Weather Rev., 98,* 106–121.
13. T. Matsuno, 1966: Quasi-geostrophic motions in the equatorial area. *J. Meteorol. Soc. Japan, 44,* 25–43.
14. R. S. Lindzen, 1970: Internal equatorial planetary scale waves in shear flow. *J. Atmos. Sci., 27,* 394–407.
15. G. D. Robinson, 1967: Some current projects for global meteorological observation and experiment. *Quart. J. Roy. Meteorol. Soc., 93,* 403–418.
16. R. H. Kraichnan, 1967: Inertial ranges in two-dimensional turbulence. *Phys. Fluids,* 1417–1423.
17. E. N. Lorenz, 1969: The predictability of flow possessing many scales of motion. *Tellus, 21,* 289–307.
18. J. Smagorinsky, 1969: Problems and promises of deterministic extended range forecasting. *Bull. Am. Meteorol. Soc., 50,* 286–311.
19. E. N. Lorenz, 1970: Climate change as a mathematical problem. *J. Appl. Meteorol., 9,* 325–329.
20. A. Arakawa, 1966: Computational design for long term numerical integration of the equations for atmospheric motion. *J. Comput. Phys., 1,* 119–143.
21. *Plan for U.S. Participation in the Global Atmospheric Research Program,* 1969: U.S. Committee for the Global Atmospheric Research Program (National Academy of Sciences, Washington, D.C.).
22. N. L. Frank, 1969: The "Inverted V" cloud pattern—an easterly wave? *Mon. Weather Rev., 97,* 130–140.
23. C. P. Chang, 1970: Westward propagating cloud patterns in the tropical Pacific as seen from time composite satellite photographs. *J. Atmos. Sci., 27,* 133–138.
24. V. Suomi, 1969: Recent developments in satellite techniques for observing and sensing the atmosphere, in *The Global Circulation of the Atmosphere* (Royal Meteorological Society, London), pp. 222–234.
25. D. Q. Wark and D. T. Hilleary, 1969: Atmospheric temperature: successful test of remote probing. *Science, 165,* 1256–1258.
26. R. M. Lhermitte, 1969: Note on the observation of small scale atmospheric turbulence by Doppler radar techniques. *Radio Sci., 4,* 1241–1246.
27. K. R. Hardy and H. Ottersten, 1969: Radar investigation of convective patterns in the clear atmosphere. *J. Atmos. Sci., 26,* 666–672.
28. E. E. Gossard, J. H. Richter, and D. Atlas, 1970: Internal waves in the atmosphere from high resolution radar measurements. *J. Geophys. Res., 75,* 3523–3536.

29. J. Cooney, 1970: Remote measurements of atmospheric water-vapor profiles using the Raman component of laser back-scatter. *J. Appl. Meteorol. 9,* 182–184.
30. C. G. Little, 1969: Acoustic methods of remote probing of the lower atmosphere, in *Atmospheric Exploration by Remote Probes, Vol. 2, Proceedings of the Scientific Meetings of the Panel on Remote Atmospheric Probing* (National Academy of Sciences–National Research Council, Washington, D.C.), pp. 619–631.
31. A. K. Blackadar and H. Tennekes, 1968: Asymptotic similarity in neutral barotropic planetary boundary layers. *J. Atmos. Sci., 25,* 1015–1020.
32. J. W. Deardorff, 1971: Parameterization of the planetary boundary layer for a general circulation model (manuscript to be published).
33. J. Z. Holland, 1970: Preliminary report on the BOMEX sea–air interaction program. *Bull. Am. Meteorol. Soc., 51,* 809–820.
34. J. G. Charney and A. Eliassen, 1964: On the growth of hurricane depressions. *J. Atmos. Sci., 21,* 68–75.
35. A. J. Faller, 1965: Large eddies in the atmospheric boundary layer and their possible role in the formation of cloud rows. *J. Atmos. Sci., 22,* 176–184.
36. R. A. Brown, 1970: A secondary flow model for the planetary boundary layer. *J. Atmos. Sci., 27,* 742–757.
37. D. K. Lilly, 1966: On the stability of Ekman boundary flow. *J. Atmos. Sci., 23,* 481–494.
38. M. Yanai, T. Maruyama, T. Nitta, and Y. Hayashi, 1968: Power spectra of large scale disturbances over the tropical Pacific. *J. Meteorol. Soc. Japan, 46,* 308–323.
39. M. Yamasaki, 1969: Large scale disturbances in the conditionally unstable atmosphere in low latitudes. *Papers Meteorol. Geophys., 20,* 289–335.
40. J. R. Holton, J. M. Wallace, and J. A. Young, 1971: On boundary layer dynamics and the ITCZ. *J. Atmos. Sci., 28* (in press).
41. Committee on Atmospheric Sciences, 1971: Weather and Climate Modification–Policies and Programs. (To be published.)
42. R. M. White, 1969: Remarks at the Dedication of the ESSA Geophysical Fluid Dynamics Laboratory, Princeton, New Jersey. *Bull. Am. Meteorol. Soc., 50,* 341–343.
43. F. G. Shuman and J. B. Hovermale, 1968: An operational six-layer primitive equation model. *J. Appl. Meteorol., 7,* 525–547.
44. G. Cressman, 1970: Public forecasting–present and future, in *A Century of Weather Progress* (American Meteorological Society, Boston, Mass.), pp. 71–77.
45. J. Smagorinsky, 1969: Numerical simulation of the global atmosphere, in *The Global Circulation of the Atmosphere* (Royal Meteorological Society, London), pp. 24–41.
46. S. Manabe, J. Smagorinsky, J. L. Holloway, and H. M. Stone, 1970: Simulated climatology of a general circulation model with a hydrological cycle, III. Effects of increased horizontal computational resolution. *Mon. Weather Rev., 98,* 175–213.
47. J. Namais, 1969: Seasonal interactions between the North Pacific Ocean and the Atmosphere during the 1960's. *Mon. Weather Rev., 97,* 173–192.

48. S. Manabe and K. Bryan, 1969: Climate calculations with a combined ocean–atmosphere model. *J. Atmos. Sci., 26,* 786–789.
49. H. Lamb, R. P. Lewis, and A. Woodroffe, 1966: Atmospheric circulation and the main climatic variables between 8000 B.C. and 0 B.C.: meteorological evidence, in *World Climate from 8000 B.C. to 0 B.C.* (Royal Meteorological Soc., London), pp. 174–215.
50. S. Manabe, J. Smagorinsky, J. L. Holloway, and R. F. Strickler, 1965: Simulated climatology of a general circulation model with a hydrologic cycle. *Mon. Weather Rev., 93,* 769–798.
51. Y. Mintz, 1968: Very long term global integration of the primitive equation of atmospheric motion: an experiment in climate simulation. *Meteorol. Monogr., 8,* 20–36.
52. S. Manabe and J. L. Holloway, 1970: Climate modification and a mathematical model of atmospheric circulation, in *A Century of Weather Progress* (American Meteorological Society, Boston, Mass.), pp. 157–164.
53. J. R. Goldsmith, 1968: Effects of air pollution on human health. *Air Pollution, 1,* 554–560.
54. M. Glasser, L. Greenberg, and F. Fields, 1967: Mortality and morbidity during a period of high levels of air pollution. *Arch. Environ. Health, 15,* 684–694.
55. J. S. Jacobson and A. C. Hill, eds., 1970: Recognition of air pollution injury to vegetation: a pictorial atlas. Informative Report No. 1 (Air Pollution Control Association, Pittsburgh, Pa.).
56. Massachusetts Institute of Technology, 1970: Man's impact on the global environment. Report of MIT Summer Study on Critical Environmental Problems.
57. R. G. Ridker, 1967: *Economic Costs of Air Pollution–Studies in Measurement* (Praeger, New York).
58. A. C. Stern, ed., 1970: Proceedings of symposium on multiple source urban diffusion models. Publ. No. AP-86 (Environmental Protection Agency, Air Pollution Control Office, Research Triangle Park, N.C.).
59. P. W. Hodge, 1971: A large decrease in the clear air transmission of the atmosphere 1.7 km. above Los Angeles. *Nature* (in press).
60. L. Kaplan, University of Chicago, 1971: Personal communication.
61. U.S. Special Commission on Weather Modification, 1966: Weather and climate modification, National Science Foundation Report No. 66-3 (Washington, D.C.).
62. H. J. Taubenfeld, 1966: Weather modification: law, controls, operations. National Science Foundation Report No. 66-7 (Washington, D.C.).
63. W. R. D. Sewell, ed., 1966: *Human Dimensions of Weather Modification* (University of Chicago Press, Chicago, Ill.).
64. R. G. Fleagle, ed., 1968: *Weather Modification: Science and Public Policy* (University of Washington Press, Seattle, Wash.).
65. H. J. Taubenfeld, ed., 1968: *Weather Modification and the Law* (Oceana Press, Dobbs Ferry, N.Y.).
66. R. W. Johnson, 1970: Federal organization for control of weather modification. *Nat. Resour. J., 10,* 222–267.

67. R. J. Davis, 1970: in *Controlling the Weather–A Study of Law and Regulatory Procedures.* H. J. Taubenfeld, ed. (Dunellen Publishing Co., New York), pp. 181–218.
68. G. Persson, 1968: The acidity and the concentration of sulfate in precipitation over Europe (The Swedish National Nature Conservancy Office, Solna, Sweden).
69. Ad Hoc Panel on Technology Assessment, Committee on Science and Public Policy, 1969: Technology: Processes of assessment and choice. Report to the U.S. House of Representatives Committee on Science and Astronautics (National Academy of Sciences, Washington, D.C.).
70. R. C. Gentry, 1970: Hurricane Debbie modification experiments. *Science, 168,* 473–475.
71. U.S. Dept. of Commerce, NOAA, 1969: Catalogue of U.S. Government meteorological research and test facilities. Federal Coordinator for Meteorological Services and Supporting Research.

PARTICIPANTS AND OTHER CONTRIBUTORS

STEERING COMMITTEE

R. G. FLEAGLE, *Chairman;* Department of Atmospheric Sciences, University of Washington

L. J. BATTAN, Institute of Atmospheric Physics, University of Arizona

G. S. BENTON, Faculty of Arts and Sciences, Johns Hopkins University

C. G. LITTLE, Wave Propagation Laboratory, Environmental Research Laboratories, National Oceanic and Atmospheric Administration

T. F. MALONE, The Graduate School, University of Connecticut

PARTICIPANTS

H. ARAKAWA, Department of Meteorology, University of California

W. R. BANDEEN, NASA Goddard Space Flight Center

L. J. BATTAN, Institute of Atmospheric Physics, University of Arizona

G. S. BENTON, Faculty of Arts and Sciences, Johns Hopkins University

J. R. BLANKENSHIP, Torrance, California

J. A. BUSINGER, Department of Atmospheric Sciences, University of Washington

R. CHARLSON, Department of Civil Engineering, University of Washington

A. S. DENNIS, Department of Atmospheric Sciences, South Dakota School of Mines and Technology

E. S. EPSTEIN, Department of Meteorology and Oceanography, University of Michigan

R. G. FLEAGLE, Department of Atmospheric Sciences, University of Washington

D. FOX, National Center for Atmospheric Research

F. A. GIFFORD, Atmospheric Turbulence and Diffusion Laboratory, National Oceanic and Atmospheric Administration

R. N. GRUBB, Space Disturbances Laboratory, Environmental Research Laboratories, National Oceanic and Atmospheric Administration
J. E. HAAS, Man-Environment Research Program, University of Colorado
T. O. HAIG, Space Science and Engineering Center, University of Wisconsin
D. S. JOHNSON, National Environmental Satellite Service, National Oceanic and Atmospheric Administration
P. R. JULIAN, National Center for Atmospheric Research
L. D. KAPLAN, Department of Geophysical Sciences, University of Chicago
W. W. KELLOGG, National Center for Atmospheric Research
R. W. KNECHT, Environmental Research Laboratories, National Oceanic and Atmospheric Administration
C. W. KREITZBERG, Department of Physics, Drexel Institute of Technology
C. B. LEOVY, Department of Atmospheric Sciences, University of Washington
D. LIPKE, Communications Satellite Corporation
C. G. LITTLE, Wave Propagation Laboratory, Environmental Research Laboratories, National Oceanic and Atmospheric Administration
E. N. LORENZ, Department of Meteorology, Massachusetts Institute of Technology
B. LUSIGNAN, Stanford Electronics Laboratories, Radioscience Laboratory, Stanford University
T. F. MALONE, The Graduate School, University of Connecticut
P. K. MUELLER, Air and Industrial Health Laboratories, California Department of Public Health
A. H. MURPHY, Department of Meteorology and Oceanography, University of Michigan
C. W. NEWTON, National Center for Atmospheric Research
H. A. PANOFSKY, Department of Meteorology, Pennsylvania State University
G. D. ROBINSON, The Center for the Environment and Man, Inc.
F. SANDERS, Department of Meteorology, Massachusetts Institute of Technology
J. SPAR, Department of Meteorology and Oceanography, New York University
P. A. SQUIRES, Desert Research Institute, University of Nevada
J. C. STRAITON, Communications Division, Weather Bureau, National Oceanic and Atmospheric Administration
V. E. SUOMI, Space Science & Engineering Center, University of Wisconsin
E. M. VAN VLECK, Ames Research Center, National Aeronautics and Space Administration
B. WEINSTOCK, Scientific Research Laboratories, Ford Motor Company
J. W. WINCHESTER, Department of Oceanography, Florida State University
G. R. HILST, *Study Director*
J. R. SIEVERS, *Executive Secretary*

GOVERNMENT AGENCY CONTRIBUTORS

E. W. BIERLY, Division of Environmental Sciences, National Science Foundation

E. BROMLEY, JR., Aviation Weather Systems Branch, Federal Aviation Administration, Department of Transportation

N. L. DUROCHER, Office of Space Science and Applications, Code SRF, Headquarters, National Aeronautics and Space Administration

R. J. ENGELMANN, Fallout Studies Branch, Division of Biology and Medicine, Atomic Energy Commission

G. HOLZWORTH, Division of Meteorology, Air Pollution Control Office, Environmental Protection Agency

C. E. MOHLER, Branch of Power Resources, Bonneville Power Administration, Department of the Interior

M. J. RUBIN, Office of Special Studies, National Oceanic and Atmospheric Administration

C. J. TODD, Office of Atmospheric Water Resources, Bureau of Reclamation, Department of the Interior

PRINCIPAL GOVERNMENT AGENCY REPRESENTATIVES

Atomic Energy Commission

R. J. ENGELMANN, Chief, Fallout Studies Branch, Division of Biology and Medicine

Department of Defense

T. H. R. O'NEILL, USN, Chief, Environmental Services Division, Office of the Director of Defense Research and Engineering

ROBERT D. FLETCHER, Deputy Chief of Staff, Aerospace Sciences, Headquarters, Air Weather Service

MURRAY H. SCHEFER, Physical Sciences Office, Naval Air Systems Command

National Oceanic and Atmospheric Administration

M. J. RUBIN, Chief, Office of Special Studies, Plans and Programs

Environmental Protection Agency

G. HOLZWORTH, Chief, Air Pollution Geophysical Research Branch, Division of Meteorology, Air Pollution Control Office

Department of the Interior

C. J. TODD, Assistant Chief, Office of Atmospheric Water Resources, Bureau of Reclamation

National Aeronautics and Space Administration

MORRIS TEPPER, Director, Meteorology Program and Deputy Director, Office of Earth Observations

National Science Foundation

E. P. TODD, Deputy Assistant Director for Research

Department of Transportation

RICHARD STROMBOTNE, Office of Physical Sciences

CONSULTANTS

J. A. CRUTCHFIELD, Department of Economics, University of Washington
M. FERENCE, JR., Scientific Research Laboratories, Ford Motor Company
M. NEIBURGER, Department of Meteorology, University of California at Los Angeles

OTHER CONTRIBUTORS

R. R. BRAHAM, JR., Department of Geophysical Sciences, University of Chicago
G. W. BRIER, Air Resources Laboratory, National Oceanic and Atmospheric Administration
C. W. HOWE, Resources for the Future, Washington, D.C.
P. C. JOBES, Department of Sociology, University of Colorado
J. D. McQUIGG, Weather Bureau, National Oceanic and Atmospheric Administration
J. M. MITCHELL, Environmental Data Service, National Oceanic and Atmospheric Administration
J. R. NEILON, Weather Bureau, National Oceanic and Atmospheric Administration
V. J. OLIVER, National Environmental Satellite Service, National Oceanic and Atmospheric Administration
R. J. REED, Department of Atmospheric Sciences, Universityof Washington
E. M. ROGERS, Department of Communications, Michigan State University
W. R. D. SEWELL, Department of Economics and Geography, University of Victoria
D. D. SMITH, Space Science and Engineering Center, University of Wisconsin
J. D. STACKPOLE, National Meteorological Center, National Oceanic and Atmospheric Administration